Unpopular cultures

Unpopular cultures

The birth of law and popular culture

Steve Redhead

Reader in Law and Popular Culture
and Co-Director of the Manchester Institute for Popular Culture,
the Manchester Metropolitan University

Manchester University Press

Manchester and New York

Distributed exclusively in the USA and Canada by St. Martin's Press

Copyright © Steve Redhead 1995

Published by Manchester University Press
Oxford Road, Manchester M13 9NR, UK
and Room 400, 175 Fifth Avenue, New York, NY 10010, USA

Distributed exclusively in the USA and Canada
by St. Martin's Press, Inc., 175 Fifth Avenue, New York,
NY 10010, USA

British Library Cataloguing-in-Publication Data
A catalogue record for this book is available from the British Library

Library of Congress Cataloging-in-Publication Data
Redhead, Steve, 1952–
    Unpopular cultures : the birth of law and popular culture / Steve
Redhead.
        p.   cm.
    Includes bibliographical references.
    ISBN 0–7190–3651–8 (hard). — ISBN 0–7190–3652–6 (pbk.)
    1. Culture and law. 2. Sociological jurisprudence. 3. Popular
culture. I. Title
    K487.C8R43   1995
    340'.115–dc20
                                                                    94–38863

ISBN 0 7190 3651 8 *hardback*
     0 7190 3652 6 *paperback*

Photoset in Linotron Sabon
by Northern Phototypesetting Co. Ltd, Bolton

Printed in Great Britain
by Biddles Ltd, Guildford and King's Lynn

Contents

Acknowledgements

My thanks to all colleagues at Manchester Institute for Popular Culture, the Manchester Metropolitan University, especially Derek Wynne and Justin O'Connor. Thanks, too, to all those around the globe who helped with the seeking out of various obscure references and texts, especially Alan Hunt, Sarah Champion, Simon Frith, Chris Stanley, Mike Featherstone, Celia Lury, Antony Easthope, Neil Duxbury, Martin Cloonan and Jon Savage. The book is dedicated to all the students who endured – and enjoyed – the undergraduate and postgraduate courses which formed the germination process of this study of the birth of law and popular culture.

Parts of section 6 first appeared in a very different version as 'Disappearing youth?' in *Theory, Culture And Society*, volume 10, part 3, 1993. Thanks to Roy Boyne and Scott Lash for first commissioning them in their original form as a review essay on 'youth culture' books.

... in the next century there will be no more books. It takes too long to read when success comes from gaining time.

Jean-François Lyotard, Preface to *The Differend* (Lyotard, 1989)

1

Introduction:
Dancing on the edge

In 'The post-modern condition: the end of politics?' in post-colonial cultural theorist Gayatri Chakravorty Spivak's book of collected essays (1990) the following conversation, transcribed from the original television debate, takes place between Gayatri Spivak and an interviewer, Geoffrey Hawthorn:

> Geoffrey Hawthorn: 'So if I understand it, then, they're not objecting to the very idea of producing narratives, they are, so to speak, dancing critically on the edge of every narrative that's produced, pointing out the silences, pointing out the unspoken, undescribed others that are implied in each of these narratives. They're not themselves concerned to put a stop to narration itself.'
>
> Gayatri Chakravorty Spivak: 'I think if one can lump Derrida and Lyotard together in this way, I think what they are noticing is that we cannot but narrate. So it's not a question of waging war on narratives, but they're realising that the impulse to narrate is not necessarily a solution to problems in the world. So what they're interested in is looking at the limits of narration, looking at narrativity, making up stories that tell us, "This is history", or making up stories that tell us, "This is the programme to bring about social justice." They're looking at that in a certain way as symptomatic of the solution. We must work with them, but there are also problems. But the other problem also is that in a narrative, as you proceed along the narrative, the narrative takes on its own impetus as it were, so that one begins to see reality as non-narrated. One begins to say that it's not a narrative, it's the way things are.'

The reasons for telling the stories in this book are, essentially, twofold. One is a desire to refute a widespread view that research and teaching at the margins, or the edges, of well-established academic disciplinary boundaries, such as law, is in some way less weighty or serious and therefore less worthwhile than scholarship supposedly concentrated at the centre, or mainstream, of such disciplines. The second reason is a conviction that sometime in the last decade the trajectories of sociology of law and sociology of deviance and cultural studies criss-crossed and therefore require a new mapping and re-orientation in order to prevent a rewriting of the history of both trajectories as if they were separate. In writing a previous book on aspects of what I had already begun – by the late 1980s – to call 'law and popular culture' (Redhead, 1990) I realised that it was necessary to try to tell a series of stories of how the regulation of popular culture came to be born, both inside and outside the academy.

The phrase 'law and popular culture', or the 'regulation of popular culture', designates both a 'disciplinary' field of study and a realm of regulation, discipline or policing, a connection which other scholars have taken from Michel Foucault in the phrase 'power/knowledge'. The phrase, and conception, follows the way that Foucault conceived it in book titles like *The Birth of the Clinic* (1975) or subtitles such as *The Birth of the Prison* (1977), though this should not all be taken to imply that any totalising 'Foucauldian' analysis is afoot in the present book. Nevertheless, Foucault's work has certainly helped to create new domains in the social sciences[1] and I, like many other academics in humanities and social studies, have found his various writings enormously helpful in the pursuit of 'archeological' work on popular culture, law, crime and deviance.

Basically, then, this book is a story about theory. As Brian McHale (1993) argues, 'where once we had theories about narrative, we begin now to have stories about theory'. There is an urgent need, in my view, to narrate the (hi)story – or indeed, more accurately, (hi)stories – of a specific regime of power/knowledge which I have labelled as 'law and popular culture'; in other words, to tell tales of its formation, scope and influence. These story tales[2] are more than merely the sum of the important scholarly work going on in the fields of the excavation of, for instance, the history of 'sumptuary laws'. Sumptuary laws are the late medieval and early modern European laws which regulated conspicuous consumption in respect of dress, ornaments, food and expenditure on the occasions of weddings, funerals and other rites of passage: they constitute regimes of regulation which some historians and sociologists of law are excavating.[3] These tales are also more than the general and various analyses, which have in fact been produced over many years, of the sphere of regulation of consumption, commodification and appearance in social life ranging all the way from the 'law creation' or 'emergence studies' of gambling and licensing laws[4] to the tales of what Paul Willis once called the 'profane culture' of the transgressions of leather-clad bike boys.[5] Rather, the histories and studies referred to in this book suggest the existence of a more widespread condition in *fin de siècle* legal and cultural theory which is inappropriately defined by the emergence over the last ten or fifteen years of labels such as 'postmodern jurisprudence' and 'postmodern sociology'.

Nevertheless, as with the attempts to forge a 'sociology of post-modernism', a 'jurisprudence of postmodernism', as section 5 of this book shows, is likely to be a difficult enterprise. The present book is, furthermore, a narrative about the history and the cultural politics of an emerging academic discipline which I have christened here, with some irony and (blank) parody, as 'popular cultural studies'. This disciplinary terrain is shown, in the various sections of the book, to emerge where 'law' and 'popular culture' meet; where the battle over terms such as 'unpopular' and 'popular', 'legal' and 'illegal', 'normal' and 'pathological', 'straight' and 'deviant' are fought out.

Part I of the book begins with the related histories of two apparently distinct approaches to the legal and social regulation of popular culture: firstly, that branch of legal studies comprising jurisprudence, sociology of law and sociology of deviance, and secondly, cultural studies. However, a book title designating *unpopular* cultures does not simply denote a study of outlawed cultures, or 'outsiders' as Howard Becker conceived them when he first asked the (still) critical question, 'whose side are you on?' in *Outsiders*, published in 1963. The question of who is an 'outsider', and the social and legal processes by which such a label is attached, remains of vital contemporary significance and, despite the legitimate criticisms made against much of its enterprise, one which labelling (or interactionist or deviancy) theory in criminology tries justifiably to this day to maintain as its object of enquiry. To decide what, and who, is 'deviant' these days – or whose side you are on now – is not an easy, or straightforward, task; in truth it was not an easy question to answer even when Howard Becker first posed it in the 1950s in his influential book, or when it became such a familiar concept for interrogation and debate in the the 1960s and 1970s. In this context it is necessary to point out that the reference to *popular culture*, in contrast to unpopular culture, in this book should not be taken to signify a celebratory inclusion of anything (echoing the so-called postmodern philosophy of 'anything goes') in the mass cultural market as a legitimate subject for analysis in 'law and popular culture'. How the terrain itself is constructed – what and who it includes and excludes – is as important as how it is approached and analysed.

The respective academic disciplines of jurisprudence, sociology of law and sociology of deviance, and cultural studies approaches to the regulation of popular culture riven by difference and conflict within their own ever-changing borders and boundaries, have themselves helped to create and sustain, as well as criticise, what we can now call a global popular culture industry. This comprises a whole international cultural realm which incorporates a whole series of hitherto disparate areas of leisure and pleasure: heritage, galleries, tourism, youth culture, the arts, fashion, the press, television, radio, information technology, theatre, film, video, sport, design and popular music *and* the cultural commentaries that go with these 'fields'. The popular culture industry has, as we approach the millennium, displaced what Frankfurt School writer Theodor Adorno critically theorised as 'the culture industry' in his writings in the mid twentieth century. Governments of various political hues have created whole departments and ministries to regulate such a rapidly expanding domain; for example, the Ministry for Culture and Communication in France and, more recently, the Department of National Heritage in the United Kingdom are seen as major sites of intervention by the state in the marketplace for popular cultural goods and services. At a local level, civil authorities have adopted similar strategies. The production of fun, or pleasure (along with their self-styled ministers and councillors 'for fun') has increasingly been given priority in an economic, legal and social sense by the state, whilst the commodification of culture (for some commentators, one possible definition of post-modernisation and postmodernism) continues apace. But if the commodification of everyday life was the focus of earlier critical theory, influenced by the Frankfurt School, the aim in this book is to take into account the aestheticisation and sexualisation of everyday life (for other commentators, the preferred definition of postmodernisation and postmodernism). To this end it provides a sketch of what an aesthetics and erotics of law might look like in the future of scholarship in 'law and popular culture'.

Essentially then, the book performs two functions. Firstly, it tells the tale of the conditions of existence of the emerging discipline which is the object of the stories – popular cultural studies – and which is in part constituted by sections of the already diverse fields of the sociology of deviance, the sociology of law, criminology, jurisprudence and various other branches of legal studies on the one hand and a whole host of versions of cultural studies on the other (or, perhaps more appropriately, the Other). Secondly, it self-reflexively sets out to extend and (de)construct the terms, contours and narrative structures of that discipline whose intended context, popular culture, is itself inherently unstable, placed as it is at an intersection of other discourses and practices. 'Popular culture', seemingly forever changing its shape at breakneck speed as an object of study, is, as I imagine it, rather like a recent post-cyberpunk novel by William Gibson[6] which was produced on computer disk and designed to self-destruct as it is read for the first time. Popular culture as the focus of an academic field of enquiry is similarly destined to unravel (and ultimately disappear) as it is deconstructed.

By way of introduction to the story of the birth of law and popular culture, the book briefly re-examines fragments of existing tales of the development of jurisprudence, sociology of law and sociology of deviance as they branched away from the narrow focus of legal studies and outwards into different fields, and of cultural studies from its roots (or routes) in English or literary studies through the social and human sciences and other approaches and back again. This is done in terms of these disciplines' institutional bases both in particular places (for instance, the Centre for Contemporary Cultural Studies at the University of Birmingham, England) and in particular countries (for example, witness the Canada/Australia/USA/Britain axis exemplified in a journal like *Cultural Studies*). The book emphasises the redesignation of the original interests and foci of jurisprudence, sociology of law and sociology of deviance and the 'relocation' of cultural studies.[7] Also, the book retells some of the stories of the body of work produced within jurisprudence, sociology of law, sociology of deviance and cultural studies from the postwar era with specific reference to the fact that these various narratives have often been apparently the product of ethnocentric, white, male (not infrequently 'masculinist') subjects or 'authors'; they have effectively been biographical, or autobiographical, in structure, although rarely acknowledged as such in any self-reflexive way in the classic texts of the disciplines. This book sees these multifarious stories as literally *his*tories of the establishment of the border/lines of new academic and journalistic enterprises. They are stories which require a radical deconstruction as well as, inevitably, more new biographical and autobiographical fragments to be added to them.

In seeking in this book to 'dance critically on the edge' of these narratives, to echo Hawthorn's conversation with Spivak which I quoted at the beginning of this section, the text also considers the policing of the discourses and practices of jurisprudence, sociology of law, sociology of deviance and cultural studies, and their partial absorption in the 1980s and 1990s into a mainstream journalism and mass media style as well as public and cultural policy. It stresses the importance of emphasising an emerging *popular* cultural studies but highlights the dangers of reading the 'popular' as always either 'conservative' or 'rebellious', or as a 'site of resistance', or as equal to, or synonymous with, the 'people'. It reflects on the resurgence of populism on both the right and the left and the new involvement of the state (at local and national government, federal and state level) in the manufacture and regulation of popular culture as a commodity. In this context, the book also relates the difficulties, and excitement, of creating a new disciplinary field in education around interdisciplinary study of the materials of popular culture, highlighting methodology such as oral history, local archival work, participant observation and ethnography, and a critical focus on questions such as the nature of fandom and fan obsession and the role of the research and the academy in celebration, regulation and mediation of 'consumer culture'.

However, what really marks out this book's review of past and present from other contemporary literature which reports from the intersections of legal and cultural theory is its central thesis. This is implied in the title of Part II of the book: namely, the disappearance of law into popular culture. Put simply, the claim is that a legal (high) modernism (defined historically as the rule of law, the legal subject, legal rights) is currently fragmenting in such a way that what could be said to be an authority, or power, to regulate and discipline the boundaries of certain social discourses and practices (for example, 'popular culture', 'postmodernism') which law once seemed to have had in modernist jurisprudential theory is now fast disappearing. The media-saturated, self-referential, culture into which law has in part disappeared (the media in many ways, as Jean Baudrillard has argued so controversially, can be regarded as a site of disappearance) presents itself for an analysis which touches on issues such as the birth of popular culture in its modern form at the end of the nineteenth and beginning of the twentieth century, its legal and social regulatory regime, the nature of rights and duties required to sustain its continuation, the reconsideration of concepts such as moral panics and law and order campaigns, and the future of the relationship between law and popular culture. The precise way in which these discussions are constituted is that relatively new 'fields' – or branches – of legal education are selected and subjected to excavation. These include sports law, entertainment law and media law as well as segments of traditional subject areas such as criminal law and property law, as well as licensing law, obscenity law and intellectual property law. The boundaries of these legal fields themselves form part of the discussion in the book of how best to understand and map the regulation of 'law and popular culture'.

The whole field of 'law and popular culture' (or law and 'play' to coin another phrase) is of increasing scholarly interest in the field of legal, social and cultural studies, not least for the massive body of regulatory instruments (court cases and statutes, local authority by-laws) now in place which require interpretation and application. In Britain, for instance, such laws seem to be, almost literally, everywhere. For example, consider the following bills: the Entertainment (Increased Penalties) Act, 1990 (dubbed the 'Bright Bill' or 'Acid House' Bill in the press) and its attacks on the organisation of what have been called pay parties or legal or illegal 'raves'; the Football Spectators Act, 1989 and its abortive compulsory identity card scheme to combat soccer hooliganism with its introduction of new measures to stop soccer fans travelling abroad and ban convicted offenders attending designated matches; the Football (Offences) Act, 1991 with its attempts to outlaw racist abuse, pitch invasions and other 'hooligan' activity at domestic soccer matches; the strengthening of licensing laws to close down certain clubs through the Licensing Act, 1988; the calls for changes in environmental and other laws to curb the noise of all-night dance parties and the 'nuisance' of the 1990s folk devils such as 'New Age travellers' and 'ravers' in various parts of town and countryside proposed in the Criminal Justice and Public Order Bill, 1993; the moral panics about Ecstasy (MDMA), LSD, cannabis and other 'recreational' (as defined by the users) drug taking amongst large swathes of late-twentieth-century 'global youth'. These regulatory regimes all exhibit familiar features of the relationships between law, market and the state in the 1990s and illustrate contemporary attempts to regulate, discipline and police popular culture in the late twentieth century which apply generally to many countries outside the national boundaries from where these specific examples are drawn. Indeed such boundaries are part of the problem, as technological and other changes make control on such border/lines almost impossible. But these aspects of legal discourse are for some commentators plainly what might be termed more or less 'repressive' in that they are seen to be part of a larger network of what many theorists persist, even in the 1990s, in calling 'social control' through criminal justice and penal systems which have in the past been theorised as part of the 'law and order control culture'.[8] In the cruder, over-simplified versions of this conception, the state, through law, is seen to be capable *only* of acting negatively – or repressively – against a group, class or individual. Power is conceived in much of this mode of theorising as a thing, an instrument, which is wielded by one group, class or individual against another. As will be seen in this book, such theorisation of legal discourse and agency is often unsatisfactory – though Foucauldian alternative theorisations of the productivity of power can be equally problematic – especially when it is focusing on new instances of folk devils,[9] moral panics or law and order campaigns.

Certainly many of the aspects of legal and cultural relations cited above do not fit neatly into the neo-Marxist theoretical frameworks which have so creatively excavated these pathways in the past. Such critical theory largely emanates from the later decades of the post war welfare states in the 1960s and 1970s and finds its objects (such as the mass, factory based, Fordist labour movement) literally disappearing before its very eyes in the 1990s. Moreover these particular features of 'law and play' are not the only instances of the increasingly complex contemporary and historical relationship between law and the whole discursive and institutional field of popular culture. In the past there has been a tendency to try to uncover this relationship either by means of some form of criminological theory (including various forms of the sociology of deviance), thereby concentrating overmuch on the criminalisation of aspects of youth and popular culture and the associated regulation and policing by various agencies including *the* police, or else by confining the analysis to 'black letter law'; that is the statutory and judge-made rules and regulations (criminal and civil) which pertain to such areas as the media, the arts, entertainment, tourism, sport and recreation. This book goes beyond such misleading dichotomies (there is, for instance, in its argument no necessary 'gap' between 'law in books', or black letter law, and 'law in action') to utilise what it sets out as diverse popular cultural studies developments and strategies in order to understand better the discourses and practices of law and popular culture. Further, it needs to be recognised that all of these assorted questions are integrally related to wider processes of internationalisation – indeed globalisation – and the transnational production, distribution and consumption of cultural goods and services, and have to be properly set in their contemporary context. For instance, new forms of conceptualisation need to be developed to cope with the swift electronic transformation of property relations which have provoked changing ideas of 'culture-as-property' in post-colonial culture. Added to this, legal and social regulation of the shifting boundaries of popular culture can be shown to be increasingly cutting across conventional (since the nineteenth century at least) divisions between public and private law, as well as state and civil society and public and private space.

Perhaps three facets are worth briefly mentioning here as the sorts of questions which might be explored in future work following the introductory telling of the stories of law and popular culture in this book. There is, firstly, the long history of law and its role in licensing popular entertainment and regulating public spaces. The changing forms of popular culture are frequently constructed, deconstructed and reconstructed through law. But where do traditional questions of legal and social regulation of popular culture within particular nation states fit into this contemporary scene? And what lasting effect will there be from changes in the rapidly shifting political and cultural geography of, say, the 'new Europe' and its manifest features including the end of the Cold War, the increase in crime, nationalism, unemployment and military conflict in the former East, and rapid changes in freedom of movement of commodities and people? Or, alternatively, what impact will the whole field of new developments in communication technologies have on legal theory and cultural politics? Secondly, the institutions of law (legal doctine, personnel, courts and so on) are integrally bound up with the changing forms of ownership and control of cultural goods and services. In Britain, the recent changes in, for example, English copyright and broadcasting law (the Copyright, Designs and Patents Act, 1988 and the Broadcasting Act, 1990 respectively) make crucial alterations for the future of many cultural industries. Thirdly, with the increase in the worldwide role of the market economy, law is becoming more heavily involved than ever before in moral censure, particularly in the domains of domesticity and sexuality – censorship laws, the age of consent for homosexual and heterosexual intercourse, 'promotion' of homosexuality by local authorities, consenting practices of sadomasochism,[10] illegitimacy, surrogacy, artificial insemination and the reorganisation of family life.

Conventionally in jurisprudential and political theory, law has been taken for granted as a 'given' – we assume that we know what it is and where to find it, and also what it does. We know, further, in this particular mythology what its objects and subjects are and what they look like – we know what and whom it regulates in advance. This picture, this book contends, is in fact a powerful (legal) fiction which may be crucial to the exercise of political power and legal authority across many different fields, especially the 'cultural'. What is necessary as part of coherent explanations for the emergence of an area of enquiry such as popular cultural studies is to explore the diverse languages of law (of law as made up of a body of texts and institutions and personnel) within a narrative and historical setting. It can, further, as we have already noted, be argued from within a modernist jurisprudence that specifically legal discourse needs to police *both* its own boundaries and the limits and contours of other discourses. The present book concentrates on the discourses and practices associated with law and popular culture and focuses on the fact that modern cultural industries are as much about 'play' as about work. This cultural politics of play, significantly, is an important element in the wider debates about the politics of popular culture in general; especially in the continuing controversies over the term postmodernism (what it means, what its uses may be), the process of postmodernisation and the condition of postmodernity. But it is also a part of a more general trend towards legal regulation of, and intervention in, leisure and pleasure as domains conceived more widely than is usually the case in traditional legal education and practice: for instance, environmental and planning law regulation of touristic uses of leisure, regulation of information technology and the illegal related practices of computer hacking and telephone phreaking, and civil liability in either tort or contract for 'loss of enjoyment'. The regulation of leisure and culture, across and within national boundaries, also tells us much about the policing – and the politics – of hedonism, pleasure, play and fun in a world which is experiencing more and more economic, social, political and environmental 'hard times'.

In one section of the book the vexed concepts and states of postmodernism and postmodernity, and their even more contentious offspring postmodernisation,[11] is subjected to re-examination in the light of legal theory, having itself finally caught up with developments elsewhere in the academy, and its usefulness for analysis of discourses such as law at one extreme, and popular culture at the other, is taken up. Without giving the 'ending' of the story away, the notion of a low (as opposed to high) modernism is preferred in this book to some of the more outrageous – and ultimately, in my view, futile – connotations of postmodernism.

The first book which I wrote for Manchester University Press, *The End-of-the-Century Party* (Redhead 1990), played jokily – and, at the same time, seriously – with the varieties of theories and signs of postmodernism which were then (mid to late 1980s) becoming current in academic and journalistic writing. Titles and subtitles of chapters stole signifiers from every conceivable (re)presentation of the 'post'. The cover photograph, by international music photographer Kevin Cummins (who also contributed the excellent inside black and white illustrations), was of what appeared to be a policeman in uniform tearing up a piece of fabric on which was printed the late 1980s familiar yellow and black smiley logo from what was then known in global youth culture (and later the tabloid press) as 'acid house'. This was, in fact, a classic piece of staged simulation. The photograph had first

appeared adorning a cover of a United Kingdom weekly popular
music paper, the *New Musical Express*, and the person performing
the ritualistic ripping up of the signs of acid house was a member of
the *NME* staff, not a real-life official agent of 'law and order'.
Furthermore, the text of the book itself was, to an extent, constructed backwards. The book opened with a 'post' script which
was in fact the last chapter I submitted to the publisher. Some
reviewers thought that its purpose was to ensure that the study of
'youth and pop' in its pages was as up to date as possible by
including research on acid house which was at the time exploding
into wider popular consciousness through extensive mass media
press coverage and police and criminal court action. I readily
concede that that may indeed have been an unintended effect.
However, one of its intended meanings was to suggest a circularity
rather than linearity – a major theoretical theme of the book – in
the writing of (pop) cultural histories; this was reinforced by
placing the original introduction of the book as the final, concluding chapter under the (plundered) title of 'The absolute non-
end'. This 'introductory' chapter was, intentionally, written at the
beginning of the project, as the book jacket 'blurb'. The effect of
this strategy was that the reader could 'know' the conclusion(s) of
the book either by reading the dust cover which contained the
distilled essence of the concluding chapter (in other words reading
the book backwards) or by reading the first chapter itself. Either
way there was no closure of the story.

In keeping with this intertextuality, and the general idea
(borrowed liberally from Jacques Derrida, theorist of deconstruction, whose work has influenced the present book pervasively)
that there is a never-ending constant 'play' of what through structural linguistics and semiotics we have come to call 'signifier' and
'signified' (or thing and referent), it is, unavoidably, the case that
parts of *Unpopular Cultures* could have appeared in *The End-of-
the-Century Party*, and vice versa, or in the reviews, essays and
books I have written in the intervening years. Where one text in any
writer's canon finishes and another one begins is in a sense highly
arbitrary; whether it is a final caving-in to editorial anxiety, or
contractual conditions, computer breakdown or just sheer
exhaustion – of finances or energy – is highly unpredictable. To say,
then, that this is a *new* book would not be entirely accurate;
new/old is yet another of the binary divisions which are decon-

structed throughout this book and are, these days it seems, always dissolving – and, also, being reconstructed – in front of us. The wide-ranging themes of the present book suggest engagement in a never-ending conversation about law and popular culture. The book is cut and pasted into sections, rather than the more fixed and less provisional notion of chapters. The penultimate section of the book is indeed both a false conclusion to this current work and a preface to other, future projects which seeks to explore anew the relations of regulation and productivity in law and popular culture discussed in these pages.

Unpopular Cultures, lastly, is also devised as a text/book on law and popular culture, a discipline in the legal and cultural studies field which is emerging in its own right on the curriculum in post-sixteen education in various countries, especially in North America where the term is increasingly common in the pages of academic law journals. In order to help readers who wish to study the relation between law and popular culture further the book contains a diverse bibliographical referencing of academic and journalistic sources and, for the same pedagogic purpose, the book is completed by a glossary of useful concepts.

Notes

1 See Gane and Johnson 1993.
2 See Papke 1991, on narrative and legal discourse in general and on story-telling and the law in particular.
3 See Hunt 1994 forthcoming.
4 See Chevigny 1991, and Redhead 1993a.
5 See Sato 1991, and Redhead 1993b.
6 The book in question is called *Agrippa (A Book of the Dead)* and is published in a limited edition of 455 copies, the cheapest at $450 and the most expensive in a bronze case at $7,500; see Edwards 1992. William Gibson's other cyberpunk output consists of *Neuromancer* (1986), *Count Zero* (1987), *Burning Chrome* (1988), *Mona Lisa Overdrive* (1989) and *Virtual Light* (1993). See also Bruce Sterling's *Mirrorshades*, (1988) for a classic anthology in this genre including an early collaboration from William Gibson and fellow cyberpunk guru Bruce Sterling. For a rather different, and in some ways more satisfying, genre than the male-dominated (and mainly masculinist) cyberpunk field, see Gibson and Sterling's collaboration on a novel, *The Difference Engine* (1990), which focuses on an imagined

Victorian world that invented the computer instead of the steam engine. For a collection of fact and fiction on cyberpunk and post-modernism, see McCaffery 1991. For discussion about whether the futuristic technology now being developed (which is of course the electronic foundation of cyberpunk's fictitious future) will make books themselves obsolete see Schofield 1993, and Fisher 1992, and note also the quotation from Jean-François Lyotard which begins this section.

7 See Taylor *et al.* 1993.

8 See Hall *et al.* 1978.

9 For the continuing relevance of such terms in the 1990s see McRobbie 1994.

10 See *R* v. *Brown, and others*, for a United Kingdom case in which the House of Lords upheld a decision of a lower court judge to convict and imprison five homosexual men for what everyone agreed were consenting sado-masochistic practices. For helpful commentaries on the case and its socio-legal context see Stanley 1993e, and Bibbings and Alldridge 1993.

11 For one account of what this concept might mean when applied to changes in advanced society, see Crook, Pakulski and Waters 1992.

Part I

Popular cultural studies

Let's talk about the future,
Now we've put the past away

Elvis Costello and The Attractions 'Less than Zero',
My Aim Is True LP, Stiff Records, 1977

2

Law and popular culture

The story of 'law and popular culture' ranges across a number of schools of thought and a number of academic disciplines. The first I wish to look at in this book are those of jurisprudence and the sociology of law.

As we mentioned in section 1, (hi)stories of disciplines are invariably founded on partial biographies and autobiographies. The stories in this book are no exception but I do seek to make the 'personal' explicit rather than, as it so often is, implicit and I subscribe, in any case, to the feminist dictum of the 1970s which proclaimed that 'the personal is political'. My own personal biography is one indication of this trajectory. After a mainly 'black letter' law undergraduate degree and a socio-legal Master's thesis, my Ph.D. work on 'law and popular culture' entitled *The Legalisation of the Professional Footballer*[1] was inspired, initially, by debates in the mid-1970s in Marxism, and other social theories, and their impact on specific bodies of knowledge such as socio-legal studies, and criminology and the sociology of deviance. I was then a full time, United Kingdom Social Science Research Council – SSRC, now ESRC – funded Ph.D. student, one of the few British postgraduates in this burgeoning global field. A decade later, after the thesis had eventually been completed part-time during my employment as a lecturer in jurisprudence, sociology of law and sociology of deviance at Manchester Polytechnic, the movement which became known as 'critical legal studies' had already made a major impact on both sides of the Atlantic, and inside and outside the academy, eventually finding its way into the 'gossip' pages of the *New Yorker* magazine in the spring of 1984 and *Time* magazine in autumn 1985 as well as taking over the entire issues of major American law journals such as, in 1984, *Stanford Law Review*.

The publication in the early 1980s of a collection of the writings of some of the early 'gurus' of critical legal studies in David Kairys's *The Politics of Law* (1982) allowed a specific evaluation to be made[2] of the international relevance of the work of contemporary critical legal theorists and of their strategic intervention in legal scholarship and the theory and politics of law-making and enforcement. Comparisons were able to be drawn between work in the USA and the frequently more theoretically sophisticated Marxist and post-Marxist legal theorising in the the UK and in Europe more generally. Marxism as a social theory, however, seemed to expose its most fatal flaws whilst it was developing a specifically legal theory in the fields of jurisprudence, sociology of law and criminology during the 1970s and, in the main, with the onset of the 'crisis of Marxism', 'post-Marxist'[3] debates, and the new theorising associated with them, have tended to eschew law as an object of study where previously it had become a central notion as major Marxist theorists of law from the beginning of the twentieth century, such as Eugenii Pashukanis, P. I. Stuchka and Karl Renner, were 'rediscovered'. In the search for what it called a 'politics of law' the Conference on Critical Legal Studies (CCLS) in the USA was an important, if uneven, resource,[4] having mushroomed in its national and international impact since its formation in 1977 and spread in particular to Europe.[5] It has to date achieved far more in the way of 'critique'[6] – Kairys's collection of essays was after all subtitled *A Progressive Critique* – than political programme or strategy, and concentrated largely on undermining, or to use its own coinage 'trashing', traditional positivist-oriented legal scholarship[7] in the academy rather than producing reform outside the law school walls.[8] Nevertheless, the terrain of debate did shift away from the initial interest in various forms of traditional and 'post-realist' scholarship[9] to analysing law in terms of its social and political effects.

What marked out the CCLS enterprise from other contemporary legal scholarship in the early 1980s was the alternative explication of legal texts which it offered, using, as Duncan Livingston (1982) points out, 'interpretive techniques of other disciplines – semiology, phenomenology and structuralism, for instance', and its stress on the importance of civil as well as criminal law, not merely its revisitation of much of the ground trodden by American legal realism before World War II. Not that this was an entirely new departure. The social history of English law inspired by writers such as E. P. Thompson (1975 and 1991) and Douglas Hay (Hay *et al.* 1975) was mushrooming at this time, and CCLS had itself already had intellectual impact abroad in the field of the 'new legal history'[10] through writers such as Morton J. Horwitz (1977). Furthermore, CCLS did manage to reverse the trend of the earlier efforts to 'contextualise' law in socio-legal studies, law and society, sociology of law and elsewhere in the 1960s and 1970s where there was concentration to an extraordinary degree on criminal law and deviance or, alternatively, on the gap between 'law in books' and 'law in action', 'unmet legal need' or legal services for the poor, rather than the substantive civil law (property, contract, tort, family, company, commercial and so on) which by far forms the greatest part of the law school curriculum. From such a narrow focus on state-defined marginal populations, and the law's relationship to them, it was not surprising that the next step was to theorise the state itself, and a massive body of theoretical work on law and the state[11] duly followed, largely from a perspective which resurrected the classic texts of Marx and Engels.[12] What was largely ignored in this move was the traditional legal philosophy or legal theory terrain of jurisprudence, which had always largely concentrated for its notions of rights, duties and capacities on civil, rather than criminal, law, and on private rather than public law. The critical legal studies movement reinstated the jurisprudential 'problematic' – legal doctrine itself – at the heart of debates in legal and cultural scholarship. The challenge to traditional positivist-based legal studies was frequently made by CCLS, then, where the defenders of the 'black letter law' faith thought they were most safe, where legal formalism continued to reign supreme. As Thomas Heller (1980) put it: 'The dominant characteristic of the present period of American legal thought is the reconstruction of a new legal conceptualism which can provide an answer to the skepticism about the value of a theory of law implicit in early legal realism.'

Whether the depth of this challenge of a new legal conceptualism was actually maintained in the 1980s and even into the 1990s, either in the United States or elsewhere, is a moot point and is largely outside the scope of this present book, but suffice it to say that by the end of the 1980s a positivist-dominated law curriculum was still the norm in most British law schools. In the United Kingdom, and other British Commonwealth jurisdictions, 'skills'-based teaching methods became the focus for battle in the 1990s between those wishing to retain a 'black letter law' approach and the 'law in context' critics of such a pedagogy as William Twining and his colleagues (Twining *et al.* 1989) showed in their new approaches to 'professional' legal education. Ironically, it was the professional bodies themselves (the Law Society and Bar Council in Britain) who provided the main dynamic for skills-oriented change in legal education as they, and their clients, tired of an outmoded positivistic law school system which continued to churn out ciphers having memorised thousands of legal rules which could now be searched out by means of information technology at the touch of a keyboard button.

Nevertheless, 'contextual' studies of law and crime persisted – as they had done since the late 1960s – irrespective of CCLS development. Weidenfeld & Nicolson's 'Law in Context' book series made major inroads into all of the traditional, black letter legal scholarship areas and the *Journal of Law and Society* (formerly the *British Journal of Law and Society*), together with American publications such as the *Law and Society Review*, have been major outlets for the dissemination of legal studies research which expressly moves away from positivism. Series such as Routledge's 'Sociology of Law and Crime' encapsulated these themes as they pertained to the late 1980s and early 1990s as did a journal such as *Social and Legal Studies: An International Journal*, which, in particular, started to feature the crossover between law and popular culture more regularly. Elements of such work had appeared in more general sociology of law and crime journals such as *Contemporary Crises* (which later became known by its earlier subtitle, *Crime, Law and Social Change*) and the *International Journal of the Sociology of Law*. In the late 1980s however, the journal *Law and Critique* was created by a younger generation of legal theory scholars, explicitly treating law and cultural relations theoretically by probing linguistic and literary theory, in particular following Jacques Derrida and Paul de Man in 'deconstruction' and Ferdinand de Saussure in semiotics (which is alternatively known as semiology or the study of signs). This latter then, significantly, was an anti-positivist linguistic and literary 'turn' rather than, as in most of the above journals and enterprises, a reliance on a broadly sociological approach to law. Legal semiotics also had its own specialist journal by the end of the 1980s, namely the *International Journal for Law and Semiotics*, produced by the same small-scale publishing venture, Deborah Charles Publications, which was in fact the imprint for law professor Bernard Jackson, whose own books in this sub-discipline – *Semiotics and Legal Theory* and *Law, Fact and Narrative Coherence* – had done much to encourage research in the field. Deborah Charles was also responsible for *Law and Critique* and further introduced innovative monographs in this new inter-section of legal studies, literature and linguistics.

What of the 'black letter law' – the corpus of legal rules – itself rather than the alternative approaches to it? Case law history across a whole series of 'traditional' and 'emerging' pedagogic areas of legal study including sports law, licensing law, intellectual property law, heritage law, privacy law, obscenity law, enter-tainments law, media law and computer law, testifies to the increas-ing importance in the law school curriculum of 'law and popular culture'. We shall examine some of these in more detail in other sections, especially section 4, but here I want to examine briefly some of the body of case law on the popular music industry, particularly as it applies in Britain, as one instance of the concrete relationship between law and popular culture in a rapidly glo-balising leisure and entertainments business. Consider cases like that, beginning in October 1993, of George Michael and his record company Sony/CBS, one of the small number of giant multi-media conglomerates now dominating international markets for a whole range of cultural goods and services where local and national regulation in terms of prohibition, or support for, music businesses or communities is becoming increasingly complicated and fraught with difficulty.[13]

This restraint of trade case[14] of George Michael, cited in law under his real name (*Georgios Panayiotou* v. *Sony Music Enter-tainment (UK) Ltd* was the court listing), was actually preceded by a number of other significant legal battles between British recording artists, composers and their music publishing and record companies over a twenty-year period. In 1974 Tony McCauley, who wrote pop chart hits of the period for Long John Baldry, The Hollies, The Foundations, Andy Williams and others, tried to end his contract – in the style of George Michael – with his music company, January Music, on the grounds that the terms of the contract with him were unfair and that McCauley had not received good independent legal advice. In a case which established a legal precedent in this area the plaintiff composer won his case in the

House of Lords. In 1988 Holly Johnson – formerly lead singer with Frankie Goes to Hollywood who had international chart success with the singles 'Relax' (which was banned after BBC Radio 1 DJ Mike Read objected to its lyrics), 'Two Tribes', and 'The Power of Love' and '*Welcome to the Pleasuredome*' LP and received 1990s royalties when all three singles were remixed and re-released – won a High Court action against ZTT, his record company, which allowed him to pursue a solo career elsewhere. The High Court judge in the case stressed that restraints in the contract were 'unreasonable'.

Apart from these contract law origins of George Michael's case, there are important copyright issues which have been litigated in Britain. We will examine this area in more jurisprudential depth in section 4, but suffice to say here that the cases of Gilbert O'Sullivan and Elton John are also relevant. In 1984, O'Sullivan, a 1970s chart singer/songwriter who recorded 'Get Down' and 'Clair' among other hits, persuaded his music business manager Gordon Mills to settle out of court after a six-year legal dispute. O'Sullivan won approximately £2 million from Mills and the full ownership of the copyright of his songs. In Elton John's case, in 1985 the singer/songwriter, along with long-time lyricist collaborator Bernie Taupin, won unpaid royalties from Dick James Music following a fifty-day court case. The judge in this case, however, would not grant the partnership duo copyright of 169 songs which were said to have made more than £200 million in world sales by that time in the singer's career though the origins of the legal dispute and its background were long and complex.[15]

In the case of George Michael, an artist who first achieved 1980s success with Andrew Ridgeley in New Pop group Wham! before pursuing a highly lucrative solo career selling fourteen million copies of his first solo LP *Faith*, the basis of the legal strategy was to follow in the footsteps of both McCauley and Johnson and extricate himself from what he regarded as an onerous record company contract. In contract law Michael's claim was, as in the other two cases in 1974 and 1988 already cited, that the contract was unfair; he claimed further that Sony Music, owned by Japanese multinational corporation Sony, was not promoting his career in the way that it should when it acquired CBS records, Michael's original company, in 1988. Michael was allowed to record only for Sony/CBS under the deal, and the company was set to make six times as much as the singer out of George Michael compact disc sales. The dispute over the contract, signed in 1988 and requiring multi-album production by the artist and a tie-up until 2003, was a major test case for the global record industry in general as well as Sony in particular[16] as, for instance, record companies could in future decide to offer single rather than multi-album deals. When Holly Johnson's own restraint of trade case went to court in 1988 the length of his tie to ZTT was a major issue. The fifteen-year tie-up of Michael to Sony/CBS could be considered by the casual observer, as well as Michael's legal team, to be of a similarly problematic nature. Although the outcome of the case, and the reasons ('ratio') for the decision are important, the analysis of such a case from a conventional, rule-bound, legalistic perspective is inadequate for a number of reasons but specifically on two grounds: firstly, that the global context of the popular music industry requires careful contextual setting if the machinations around such disputes based on contractual and intellectual property are to be properly understood, and, secondly, that the globalisation (as well as localisation) taking place in such cultural industries makes 'national' regulation through legal rules passed in one nation state virtually impossible to predict with any reasonable certainty.

As we have seen in this section, academic history of jurisprudence and sociology of law, as well as case law history, is one of the overall intellectual contexts out of which 'law and popular culture' emerged. In section 3 we will look at the cultural studies and sociology of deviance antecedents. For the present we need to consider what studies and histories have been produced in 'law and popular culture' which follow in the traces of jurisprudence and sociology of law. Book ventures in the specific field of 'law and popular culture' have been relatively few and far between. Let us refer to one or two texts in particular. One example is the work of law and sociology professor Alan Hunt, for a long period a relatively lone pioneer in theorising in the field of law and society, sociology of law and Marxist theory of law, who has painstakingly researched the field for a book entitled *The Governance of the Consuming Passions: A History of Sumptuary Law* (1994) on legality and the long historical origins of 'modernity' and popular culture, an area of study we have already noted in section 1. Hunt's fascinating and detailed study looks at the history of 'sumptuary laws', that body of law regulating conspicuous consumption which dates back to classical civilisations but which in fact lasted until late medieval and early modern times in Europe.[17] As part of what he intriguingly refers to as a 'sociology of governance', Hunt argues that 'in the modern world the regulation of consumption persists, but becomes dispersed throughout a range of both public and private forms of governance'. He concludes that far from the 'death of sumptuary laws' having occurred in the seventeenth century, there is a persistence of projects of governance of 'personal appearance and of private consumption' into and through the next three centuries. As we shall see in section 3 of this book, the project of governance that Hunt identifies as one of 'regulation of popular recreation' can and should be explored in other contexts. Hunt's purpose, though, in tracing the history of sumptuary laws is not just to explore the obscure social history of legislation for its own sake. He wishes, as with his earlier *Explorations in Law and Society* (1993) to produce more general theorising of legality and its historical formation, and the aim of a history of sumptuary laws is partly to show that 'studies of legal forms of regulation need to be situated within the wider projects of governance, that is, the economic, social and moral regulation in which legal regulation is embedded.'

The theoretical underpinning of this theory of governance is the work not of Karl Marx (or Max Weber for that matter) which had governed Hunt's earlier published volumes but of Michel Foucault, whose method of prolonged and detailed archelogical excavations Hunt has sustained quite remarkably in his study of sumptuary laws and the 'birth' of consumption. Hunt has summarised Foucault's effects in terms of his theoretical and methodological influence on the domain of jurisprudential theory[18] but it is worth mentioning at this juncture the popular cultural context and impulse of Foucault's own 'digs' into the complexities of modernity and 'juridification'. As American academic James Miller's riveting biography of Foucault (1993) makes crystal clear, Foucault's involvement in the experience of taking LSD ('acid') and his obsessive penchant for the culture of American gay bath houses, at least since the mid-1970s, gave an autobiographical context to Foucault's apparently archaic, and at times unfathomable, search for hidden connections between law, culture and the 'self' as sexual subject. It is significant that Miller pursues an 'other', non-academic career as a rock music culture writer (for *Rolling Stone* amongst other publications); this explains, perhaps, why the popular cultural context of Foucault's archeologies of knowledge/power have been brought to the fore only in Miller's study.

Another noteworthy book-length project in the field of 'law and and popular culture' is Paul Chevigny's *Gigs* (1991) which, also referring to and reflecting the work of Michel Foucault, studies the socio-historical conditions for law and the consuming subject. This time it is jazz and the cabaret laws in New York city, as the subtitle of the book has it, that is the focus. Published in a Routledge series enticingly entitled 'After the Law', this New York University law professor's book narrates the story of the 'end of the cabaret laws' in that city, a story moreover in which Chevigny played no small part. He, along with other lawyers, worked as a volunteer on a case in the late 1980s, supporting the musicians' union, which attacked the local regulations ('the cabaret laws') that restricted the playing of live music in bars and restaurants in New York City from 1926 until 1990. Much of the book relates how the laws imposed a complex system of licensing combined with zoning restrictions on neighbourhoods where live music could be played and how, despite periodic pressure group activity from the (mainly jazz) musicians involved, the discriminatory system remained in place for most of the twentieth century. Equally the book tries to answer the pertinent questions for a 'politics of law': for instance, why did this particular case, which Chevigny supported so strongly, succeed and provoke the court into declaring the regulations of New York City unconstitutional, leading to their complete overhaul? Thus far the project, as I have described it, might seem to be a conventional socio-legal study, containing as it does discussions of the use of law and society approaches in the text of the book. However, as I have emphasised, it also includes commentary on the usefulness, or otherwise, of Foucault's theoretical histories of 'law and discipline' and, moreover, Chevigny's book goes deeper into 'law and popular culture' than more conventional 'gap' problem studies and points to elements which might distinguish socio-legal studies from 'law and popular culture' as a field in its own right. His account ranges across the history of jazz – a part of popular culture which has a special attraction for Chevigny who excavates from the position of a participant observer, a genuine fan – as well as across debates about planning, arts and culture, regeneration and the (post)modern city. There is a most useful appendix showing a chart of the New York City Cabaret Licensing and Zoning system from 1916 to 1990 and a note on book sources as diverse as Samuel Charters, Walter Benjamin and Richard Sennett. It is, though, most significantly a book which helps to contribute to research and policy and at the same time takes popular culture seriously, especially the urban night-time economy and its regulation. I discovered its useful functions in my own research work,[19] which, although concentrating on a different historical period in a different city and country (1980s and 1990s in Manchester, England) and a form of music (house, and other contemporary dance music cultures) far removed from jazz, benefited enormously from a reading of Chevigny's project.

What essentially marks out the 'modern' terrain on which I have been 'dancing on the edge' in this section is fandom. Studies of fan obsession, and their legal consequences, have also been prominent in the study of 'law and popular culture'. Fan obsession, as Fred and Judy Vermorel (1985 and 1989) have pointed out, is sometimes terminal, for either the object of desire or else the fans themselves. This can be seen in the case of John Hinckley, who attempted to assassinate the (then) President of the United States, Ronald Reagan, in 1980. His obsession with the actress Jodie Foster – and especially her performance in the Martin Scorsese film *Taxi Driver* starring Robert de Niro as anti-hero Travis Bickle, a character who stalked a political candidate and was prepared to assassinate the politician until he was scared off – has been analysed by Roseanne Kennedy (1992) in terms of the 'hyper-reality' of American media and popular culture and the consequences for the eventual court trial of Hinckley. Hinckley was declared insane at his trial, a decision which depended partly on the use (including the screening) by his defence team of the film *Taxi Driver* in the court room. Another similarly celebrated and analysed legal case was that of Mark Chapman, who assassinated former Beatle John Lennon. An identification with Holden Caulfield, the sixteen-year-old delinquent pursuer of 'phonies' in J. D. Salinger's *The Catcher in the Rye*, first published in 1951, was seen by many researchers[20] and commentators, including Chapman's defence team, as one of the plausible explanations for Chapman's crime, to which he later pleaded guilty, thereby deflecting further legal debate about his sanity. Certainly, Chapman was shown to have bought a copy of the book on the day he committed the crime, and to have studied another copy for a long period before the killing.

There is, besides specific studies such as those of Hunt, Chevigny and Kennedy which have been briefly analysed in this section, a more general interrelationship between popular culture and the 'high culture' theorising of jurisprudence and sociology of law (and other disciplines) in the academy. The connection between sociological and jurisprudential study of law and crime and popular music culture, for instance, is in fact wide and diverse. Witness at one ephemeral level the quotation from pop songs at the beginning of erudite academic essays, books and papers. A few random instances will suffice to illustrate the point. Taylor, Walton and Young's *The New Criminology: For a Social Theory of Deviance* (1973) is prefaced by a quotation from Bob Dylan's 'Absolutely sweet Marie' on the 1966 double LP *Blonde On Blonde* which begins 'To live outside the law you must be honest', and their edited collection of essays published as *Critical Criminology* (1975) begins with a 'Dedication' which comprised a verse from Dylan's 'All along the watchtower' on the 1968 *John Wesley Harding* LP that includes the line 'The thief he kindly spoke'. The same 'Thief' appears in another verse quoted from the same Bob Dylan song in John Mepham's seminal contribution to the mid-1970s search for a Marxist theory of ideology in issue number 6 of *Working Papers in Cultural Studies*, (as we shall see in section 3 at that time the house journal of the University of Birmingham Centre for Contemporary Cultural Studies). The verse reads: 'There must be some way out of here, Said the Joker to the Thief, There's too much confusion, I can't get no relief'.

In this type of connection between pop music culture and the sociology of law and crime the technique is obviously to plunder popular songs for references to 'law and crime' in order to provide a relevant prefatory quotation and the intention signalled is a desire to be seen to be aware of popular cultural developments whilst retaining an academic, high culture position from which to write. The late 1960s counter-cultural influence of rock music culture, and an all-embracing 'rock ideology', ensured that the examples I have quoted at random were merely the tip of an iceberg of 'law and popular culture' connections in this vein. Nevertheless, for a younger generation of scholars, these kinds of self-consciously referenced links persisted, though inevitably with different artists cited. For instance, a similar strategy of plunder is adopted in the debate between Chris Stanley and Peter Rush in *Law and Critique* in 1991 when in a debate over styles of legal education the two authors adopt the title 'Killing me softly with his words' from a Roberta Flack song, as a partial title for both authors' separate contributions on what the contemporary law student might be expected to gain from the academic legal curriculum. Significantly, though, Stanley also used a graphic of late 1970s 'post-punk' Mancunian band Joy Division's 'Love will tear us apart' single as a logo for his materials for a legal theory course handout for his undergraduate students at the University of Kent at Canterbury in the 1990s. The 'post-punk' sensibility, distinguishing itself radically from counter-cultural notions of an earlier era, comes through in an exciting and innovative course whose eclectic material ranges from complex European social theory to the screening of popular cult films like Ridley Scott's *Blade Runner (The Director's Cut)*. In addition to this widespread use of popular rock music and film culture, Chris Stanley (1993b) and other legal theorists such as Ian Ward (1993) have explored the links between law and 'high' and 'low' literature, a link betwen law, deviance and popular culture which we discuss extensively in section 6 of this book.

A further and more extensive use of this plunder – a widespread contemporary cultural practice in disciplines other than legal and sociological theory in any case – is evident in the various essays in the *New Mexico Law Review* 1986 special issue on the teaching of law through rock music. As Gary Minda points out (1986) in his own essay in the journal, two law professors at the University of New Mexico Law School, Karl Johnson and Ann Scales, select Tina Turner's song 'Steel claw' (a term which is coupled with the phrase 'cold law' as an image of the law) from the 1984 LP *Private Dancer* as a narrative about 'everyday experiences which run counter to the experiences of social life discussed in law', and proceed to justify its use in legal education alongside other artefacts from contemporary popular music culture, such as Woody Guthrie's 'This land is your land'. Here the argument is essentially that 'rock and roll' culture is a useful – in some cases indispensable – medium for talking about law to law students in the academy of the late twentieth century. Traditional methods of legal education have failed, runs the argument, so use of a non-legal medium of communication should displace, or add, to them.

Other American legal academics such as Jennifer Jaff (1986) and Anthony Chase (1986a) have reacted in similar fashion to the idea that music and other art forms of popular culture can be useful for understanding the everyday experiences of law, lawyers and the legal system shared by ordinary persons in modern American culture. Jaff plunders Joni Mitchell's 1985 *Dog Eat Dog* LP as well as 'Johnny 99' from Bruce Springsteen's 1982 *Nebraska* LP amongst other examples of contemporary popular song in a trawl through references to law and lawyers in popular music which ends with 'namechecks' for Jackson Browne's *Lawyers in Love* LP from 1983 and Warren Zevon's track 'Lawyers, guns and money' from the 1978 *Excitable Boy* LP.

The problem with these citations is their reflection of a mainstream (mainly white) American rock culture which is far from radical in its challenge to contemporary culture; the mythology of rock music as a 1990s counter-culture is reinforced by continually using these kinds of references and citations as a 'critical resource' for teaching legal and cultural theory. As we shall see in section 4, Houston A. Baker Jr (1993) makes it clear that 'black studies' within the American academy can utilise rap music, for instance, as a pedagogic tool through many different disciplines, including legal and literary studies, but it has to careful when so much of the musical form reinforces prejudice and stereotypes devoloped in mainstream popular culture.

Anthony Chase goes further in his steps toward a legal theory of popular culture in an article which ranges from problems with the politics of the critical legal studies movement to popular culture references to law and lawyers in film and literature as well as popular music. In the course of his own list of connections, and his extensive referencing of other attempts, Chase poses the question of why law and popular culture has remained unexamined by legal scholars. Why, he muses, has there been so little excavation from legal studies of the field of what he terms 'popular culture studies'? He also argues (1986b) in the course of a review symposium on law and culture for the American Bar Foundation that what he perceives as a gap in the research is particularly surprising in view of the fact that 'pictures of the legal system have been communicated extensively through the 'formats' of popular culture (e.g. movies, best sellers, soap operas, television news, advertising, pop music, jokes and stand-up comedy routines)'. In America, at least – and by implication other Western-style parliamentary democracies with well established legal professions – Chase feels that the study of 'law and popular culture' would be fruitful. He says that:

> the investigation into popular culture formats reveal mass media/ mass society attitudes or 'structures of feeling' regarding American law and the legal profession which can certainly help us to develop a sharper focus on what Americans really think about law and how the system within which they operate really works. A step beyond the mere exegetical tradition of doctrinal legal scholarship (which for decades passed itself off as authentic social thought) is patently required. The study of law and popular culture seems to me to be one promising avenue of original inquiry.

Jerry Frug, Professor of Law at Harvard Law School, broadens this question out in an essay in the *New York Times* (1986) where he tries to answer the query 'why should a law professor be interested in literary theory?' Both Chase and Frug imply that 'law and popular culture' can help to break down the dominance of a positivist-oriented, 'black letter' or 'exegitical' tradition of legal studies teaching. This is certainly a fruitful line of inquiry but the the argument in the remainder of this book is essentially that to get 'towards a legal theory of popular culture' we also have to develop a popular cultural theory of law – a ('low modernist') aesthetics and erotics of law as it is described in sections 4 and 5. Academically and professionally it is patently difficult to maintain expertise in both legal theory and cultural studies, but it may be necessary, not just simply to break down the high/low culture divide which is so debiliating inside the academy but because cultural studies impinges on jurisprudential thinking as never before[21] so it becomes easier than it once was to argue in legal education debates that a legal theorist must also be a (popular) cultural theorist.

Notes

1. The full title is *The Legalisation of the Professional Footballer: A Study of Some Aspects of the Legal Status and Employment Conditions of Association Football Players in England and Wales From the Late Nineteenth Century to the Present Day* (Ph.D. thesis, School of Law, University of Warwick, 1984). The thesis, regrettably, was never published. Publishers in Britain in the mid-1980s seemed to be interested in soccer culture only in terms of 'hooliganism', violence

and law-and-order debates (a situation which only altered in the early 1990s), and though these were interrogated in the thesis, reference to them was oblique.

2 See Redhead 1984, for one account. The book proved to be an inspiration to younger scholars in other countries. For example, some writers were later to take some of the essays in *The Politics of Law* out of their United States context and specifically apply them to United Kingdom conditions; Chris Stanley, for instance, published an article on British legal education in *The Law Teacher* almost a decade later under the title 'Training for the Hierarchy', paying due homage to Duncan Kennedy's seminal article of the same name in Kairys's book.

3 The main protagonists in the 'post-Marxism' debate (at least, circa 1987, in the pages of *New Left Review*) were Norman Geras on the one side, and Ernesto Laclau and Chantal Mouffe on the other. All concentrated on the political and the social rather than the specifically 'legal' instance. For the renewed Marxist theory of law debate in the 1970s and early 1980s, see Bob Jessop's summation (1980), Roger Cotterrell's update and critique (1981) and my own tentative contributions: an essay in *Critique* (Redhead 1978) and a longer overview in an excellent edited collection on *Marxism and Law*, published in the United States (Beirne and Quinney 1982). The context of these debates was frequently the revisiting of historical developments in Stalinist Soviet society and legal and political constitution (see Beirne and Hunt 1988, and Sharlet and Beirne 1984), and Nazi, and post-Nazi, German society and legal and political constitution (see Tribe 1981). For the effect of the 'crisis of Marxism' on critical legal studies in general, see Journes 1982. For work which did start to consider the 'legal' instance anew once the Marxist theory of law debates had died away see Fitzpatrick 1983, Fitzpatrick 1992, and Hirst 1988.

4 For one perspective from one of the movement's leading theorists, see Unger 1983, and for a contrasting, later, critical view from a literary theorist see Fish 1987. For an assessment of Unger's importance to critical legal theory, see Collins 1987. On literary theory and critical legal theory in general, including the contribution of Fish, see Norris 1988.

5 See Fitzpatrick and Hunt 1988.

6 See, for contrasting views on the value of 'critique', Hunt 1987, and Hirst and Jones 1987.

7 See Tushnet 1981, and Freeman 1981.

8 See Livingston 1982.

9 See Tushnet 1980.

10 See especially Hunt 1986, and also Snyder and Hay 1987, and Rubin and Sugarman 1984.

11 See, again, Jessop, 1980 and my own review of Bob Fine's *Democracy and the Rule of Law* (Redhead 1985).
12 See Cain and Hunt 1979.
13 As Simon Frith, Tony Bennett, Lawrence Grossberg and others have examined in considerable detail; see Bennett *et al.* 1993.
14 See Wittstock 1993.
15 See Philip Norman's comprehensive book on Elton John for more legal and biographical background to the case (Norman 1991).
16 See Cope 1993.
17 Historians such as Peter Burke, amongst others, have traced some of the contours of this general field of the origins of modern popular culture but in nothing like the legal and contextual detail produced by Alan Hunt.
18 See Hunt and Wickham 1995 forthcoming.
19 Supported by the Polytechnic and Colleges Funding Council (PCFC) in the United Kingdom, and conducted with colleagues at the Manchester Institute for Popular Culture, the Manchester Metropolitan University. Where I have drawn on aspects of our research in the present book, I would like especially to acknowledge the research assistance of Andy Lovatt.
20 See Bresler 1989, and Jones 1992.
21 See the essays in Fitzpatrick 1991.

3

Cultural criminology

The development of cultural studies over the past half century has in a sense been essentially a tale of two disciplines – English studies and criminology, or more accurately literary theory on one side and the sociology of deviance on the other. Though they appear distinct, the rather unorthodox genealogy of their emergence as disciplines which I develop in this and other sections in the present book shows a considerable overlap. As I have argued elsewhere[1] the forays into youth culture and youth subcultures by the Centre for Contemporary Cultural Studies (CCCS) at the University of Birmingham, England, in the 1970s partly grew out of research into 'politics and deviance' (as one of their book titles proclaimed it) by the National Deviancy Conference in the late 1960s and early 1970s. There is, too, something of a hidden history of cultural studies as a field, or even as a variegated discipline, in this archeology of recent 'criminological' and 'cultural'[2] knowledge as well as a more general intellectual history of political and academic moments in the 1970s and 1980s.

Let us first look at some of the minutiae of this history. To some extent this historical convergence represents an overlap of personalities. Stuart Hall, the director of the CCCS in its most influential period, delivered papers at the National Deviancy Conference Fifth Symposium in April 1970 and the Twelfth Symposium in January 1973, whilst Phil Cohen, perhaps the other most influential and seminal theorist in this period of CCCS work, presented a paper on 'Youth subcultures in Britain' at the Sixth Symposium in October 1970. Hall, in this area, was to be a legendary guru for much of the 1970s and 1980s and scholars in the field of cultural criminology, as well as many other academic and political domains, owe him an enormous debt. Deviancy writers, equally, had input into cultural studies publications. For instance, an early issue of the house journal of the Centre for Contemporary Cultural Studies, *Working Papers in Cultural Studies*, included in its pages an essay by deviancy theorist Stan Cohen, (1974), whose crucial publications on theories and studies of deviance, namely *Folk Devils and Moral Panics* and a Penguin paperback book edited for the National Deviancy Conference entitled *Images of Deviance*, had already been published. In an insightful author's note at the end of the essay Stan Cohen explained that:

> This paper was originally prepared for a book on youth work and leisure and aimed primarily at an audience of youth workers. This accounts for the style: unacademic, somewhat polemical and more than a little thin conceptually. It was written at the beginning of 1971 and this in a subject area such as youth culture, accounts for it being so touchingly out of date. The *Working Papers'* editors have convinced me that it is worth publishing completely unchanged (only the references have been up-dated) and to rescue it from the status of a quaint historical relic have allowed me to make these few observations. They are confined to problems other than those of simply incorporating developments in contemporary pop culture over the last three years, although this is no easy task in the light of phenomena as diverse as David Bowie, the Osmonds, Alice Cooper and the extraordinary difficulty now of finding any sort of identity in the current stagnation of pop culture ... a related point, one of considerable theoretical density and not just a matter of 'taking into account' further research, is connected with the current work at the Centre for Contemporary Cultural Studies on the development of youth cultures in post-war Britain. My paper wholly glosses over the complex link between history and subjective experience or (more concretely in this case) the links on the one side between the history of youth subcultures and their articulation in the dominant culture and on the other, their intrusion into the individuals' biography. It remains to be seen whether current work on skinheads, Teddy Boys, hippies and the like can do justice to both these forms of analysis.

It was not, however, merely an overlap of personnel which made the convergence of sociology of deviance and an emerging cultural studies worthy of remark, nor even the pre-punk note of disillusionment with the state of 'pop'. As Stan Cohen hinted in this passage, the theoretical convergence, even where it highlighted problems in earlier theories of deviance, including his own, was vital. He went on in the author's note quoted above to say that:

The work currently being carried out by Laurie Taylor and myself in this area has tried to be much more careful than my 'Breaking Out' paper is in excavating the meaning of individual actors' statements of their own motivation. Although I would want to retain the emphasis on showing how society only allows what I called the 'more glamorous deviant' an ideological meaning, we are somewhat more sceptical of our earlier attempts in this field which might have led to the spurious attribution of such qualities. The reverse problem is also apparent: in trying to normalise forms of deviance by rescuing them from the clutches of positivist criminology and the grosser stereotypes of the media and control agents, one might miss those cases (and certain forms of breaking out and smashing up are included) in which the rejection of eveyday life is more noteworthy than the institutionalised, almost banal, features of the deviance on which I laid such great stress.

The same issue of the *Working Papers in Cultural Studies* journal which contained Stan Cohen's article publicised the *20 Years* pamphlet which covered a recent notorious 'mugging' case in Handsworth (an inner-city area of Birmingham); the label 'mugging' had, as the pamphlet and later CCCS research noted, recently been 'imported' from its more regular use to describe 'robbery with violence in the open air' on the streets of the USA. The fanzine graphic-style front cover of the pamphlet predated mid-late 1970s punk fanzines as well as recalling the early 1970s anarchistic radical lawyers' magazine, *Up Against the Law*. The back cover of the pamphlet (written and published in 1973) itself acknowledged the help of *both* the members of 'the Centre for Contemporary Cultural Studies, University of Birmingham, and Stan Cohen'. CCCS was praised for its 'invaluable help with research and design'. CCCS were also quoted as 'wishing to acknowledge their debt to the theoretical work of Stan Cohen of the University of Essex'. The *20 Years* pamphlet was, in reality, part of the CCCS ongoing research project on law and order, mugging and the state. The eventual book text was published in 1978 by Macmillan[3] and credited to (then) CCCS members Stuart Hall, Chas Critcher, Tony Jefferson, John Clarke and Brian Roberts. The final book version – itself over 400 large pages – was a slimmed-down version of a much larger manuscript which had circulated in reading groups of academics interested in these questions for some time before that. By the time the book appeared, the National

Deviancy Conference had fragmented considerably, but the 'Acknowledgements' page still sought to point to the CCCS 'mugging' project's strong, original, roots in the NDC. As well as thanking prominent individual NDC members such as Stan Cohen, Ian Taylor, Mike Fitzgerald and Jock Young for their 'detailed comments and care' the authors noted that:

> Although we have borrowed ideas and concepts and worked them in directions which they may not altogether approve, we have had nothing but positive encouragement and support from those people in particular, and the context in which those conversations first arose – that of the National Deviancy Conference.

The vitally important CCCS work of writers such as Paul Willis, John Clarke and Dick Hebdige was also based around the main sociology of deviance concerns of the era, as was in later years that of Paul Gilroy, who eventually wrote much more extensively on black cultural politics, the terrain on which the 'mugging' pamphlet had been prepared. In the late 1970s and early 1980s Gilroy contributed incisive and provocative essays from within the CCCS working group concentrating on 'race and crime'. One contribution in particular, named 'police and thieves' after a legendary Junior Murvin reggae track much beloved of 1970s punks as well as black music fans and DJs like Gilroy, was included in *The Empire Strikes Back* CCCS collection published in the early 1980s, whose own title wittily parodied the name of a sequel to *Star Wars*, a massively popular science fiction film of the time. The essay put Gilroy squarely in the camp of the 'left idealist' group of criminologists and in the vanguard of opposition to those, like Ian Taylor and Jock Young, who were perceived to have moved away from the radical deviancy and 'new criminology' theory of yesteryear and who eventually became dubbed with the misleading epithet 'left realists'. Gilroy's thesis on 'police and thieves' was prominent in the arguments since law, policing and young black street crime was the main ground on which this vitriolic, often sterile, criminological debate[4] was fought out in the 1980s, though this had to be situated in popular music culture, as sections of his later book *There Ain't No Black in the Union Jack* made clear. Moving away from these initial deviancy concerns in the 1990s Paul Gilroy[5] has produced important, though still controversial, forays into black nationalism and popular culture. From the late

1970s onwards, Dick Hebdige has continued to develop and revisit the youth and deviance couplet. Hebdige's own MA thesis from the Centre for Contemporary Cultural Studies was entitled *Aspects of Style in Deviant Sub-cultures of the 1960s* and formed the basis for four (nos 20, 21, 24 and 25) of the CCCS Stencilled Papers series. Both organised crime and youth delinquency were focused upon in paper no. 21, *The Kray Twins: Study of a System of Closure*, where he combined the two themes in a provocative analysis of the celebrated late 1960s film *Performance* featuring Mick Jagger, Anita Pallenberg and James Fox. In Stencilled Paper no. 42, *Football Hooliganism and the Skinheads* John Clarke developed his earlier work on skinheads and working-class culture in a specific critique of the radical deviancy theory and new criminology theory work of Ian Taylor on embourgeoisement and football hooliganism (Taylor 1971). One of Paul Willis's two major book projects of the 1970s, *Learning to Labour* and *Profane Culture*, was eventually made up of the development of Stencilled Papers on 'bike boys' and 'hippies' published by the CCCS. His beautifully written study of 'lads' culture' in 'How working class lads get working class jobs' also focused on the issues of truancy and delinquency in and out of school by young working-class males. Perhaps Willis's work, too, has remained most significant for its exemplary ethnographic, participant observation emphasis which could today be seen as CCCS's permanent monument to the study of youth, deviance and culture.

To some extent the convergence of cultural studies and deviancy theory in the early to mid 1970s became a divergence by the end of the decade. This was becoming obvious, however, as early as 1974–5. In summer 1975 CCCS produced its most long-lasting issue of *Working Papers in Cultural Studies*. A double issue, eventually formally published in 1978 by Hutchinson as a book edited by Stuart Hall and Tony Jefferson, WPCS no. 7/8 remains a fascinating social and theoretical document. Although much interest in subsequent years has justifiably centred on the 'male-ness' (authors and subjects) of many of the ethnographies – a subject which was precisely the object of critique in the contribution by Angela McRobbie and Jenny Garber in the collection itself – the main dissenting essay, entitled 'The politics of youth culture' was in fact written from outside the CCCS membeship by Simon Frith and Paul Corrigan. Corrigan and Frith were then lecturers in the Department of Sociology at the nearby University of Warwick at Coventry: both were at the forefront of this field and keenly interested in the development of sociological work on youth and education. Along with other members of the Warwick University staff interested in the sociology of deviance and 'social problems' research, such as Bob Fine, Frith and Corrigan actually met in research meetings with the CCCS group from Birmingham and found a considerable theoretical gap between the two sides; a greater distance certainly than the twenty miles which separated the two institutions!

This gap surfaced in later debates – such as book reviews – as the main bones of the contention became clear: that however interesting and innovative the theoretical melange of interactionism, phenomenology and Western Marxism in the body of CCCS literature on youth culture, the empirical evidence for many of the youth subcultural contentions was either highly debatable or else simply absent. Debate centred, too, on the 'autobiographical' roots/routes of the CCCS work; youth subcultures such as mods from the 1960s were seen as an important generational formation for the work of Dick Hebdige for instance. Simon Frith went on to produce an impressive international ouevre of writings in the field of what came to be seen as 'popular music studies'. Almost single-handedly and at considerable 'academic' cost in the early days,[6] Frith created an international academic disciplinary field of popular music studies, originally emanating from his own pioneering book *The Sociology of Rock*, first published in 1978, which was itself later completely rewritten to constitute the seminal *Sound Effects*, first published in 1981 in the USA. Paul Corrigan's own major contribution to the debate on youth culture and deviance was on the theme which he termed the 'dialectics of doing nothing' and featured his idiosyncratic[7] ethnography of working-class ('smash street') kids in the north east of England. The book – which was the published version of his Ph.D. thesis at the University of Durham under the supervision of Stan Cohen – was already in preparation when the debate with CCCS began but owing to various mishaps (including the burning down of a printing press!) publication was delayed until 1979 (Corrigan 1979).

Meanwhile, the National Deviancy Conference had resurfaced. It held a major conference in 1977 on the theme of 'Whatever happened to the sixties legislation?', contributions on which were published several years later as an edited book on 'permissiveness and control' (NDC 1980), with Stuart Hall again a prominent contributor and editor, along with other CCCS members such as John Clarke. By the time the book of essays was published, the National Deviancy Conference had moved into yet another phase by becoming entwined with the Conference of Socialist Economists (CSE) and in particular its 'Law and state' group. There was also, incidentally, a pamphlet of the same name, *Law and State*, which ran to several issues, and, although it was modelled on the radical law magazine *Up Against the Law*, it was only loosely connected to the CSE and was organised by the Glasgow East End Law group. The fruits of this CSE/NDC collaboration were seen in a jointly edited book[8] based on papers from an earlier joint conference, involving individuals from both 'camps' such as Bob Fine on the one hand and Jock Young on the other, and proclaiming in its subtitle the theoretical and political transition 'from deviancy theory to Marxism'. The collection included a theoretical essay by Phil Cohen which combined his renewed interest in 'policing the working class city'[9] with reflections on his earlier formative studies of youth subcultural theory. Both Cohen and his collaborator, Dave Robins[10] continued to follow these theoretical and political interests into the 1980s and 1990s with special focus on football and youth culture.

It is my contention in this book that the original concerns of 'deviance' theory of the 1950s and 1960s (transgression, rule breaking, the 'pleasures' and 'pain' or transgression) have been kept alive, and then resuscitated, within cultural studies in the 1970s and 1980s. What I will designate in this book as popular cultural studies refocuses this theoretical interest which in the 'literary' and 'postmodern' turns within cultural studies has tended to become buried, or pushed into blind alleys. It is sociology of deviance (and to some extent, as we saw in section 1, legal theory in jurisprudence and sociology of law) which put the 'popular' into cultural studies by reintroducing the focus on the social reaction to the act of rule breaking, or transgression. Teaching and researching youth, or popular, culture became the natural outcome of a twenty-year trajectory for many scholars of the 1970s and their subsequent students. A move, then, out of 'deviancy studies' and into various, fragmented cultural studies fields became an identifiable academic career pattern. This trajectory, crucially, came to include the generic area of 'leisure studies', which spawned its own journal *Leisure Studies* and a newsletter and association (the Leisure Studies Association) where many of the original University of Birmingham cultural studies writers (such as John Clarke and Chas Critcher, amongst others, who went on to co-author their own leisure studies textbook *The Devil Makes Work*) published their subsequent writing and found international networks and new centres.[11] Indeed a field such as leisure studies in the 1980s was theoretically similar to deviancy studies in the 1960s in that it for many years often exhibited the familiar mismatch of theoretical knowledge and empirical research. Some writers have moved more directly – though still over a number of years – from practitioners of the labelling theory of deviance into high priests of postmodern theory. For example, Stephen Pfohl (1992) and Norman Denzin (1991) have both written fascinating, though very different, books in the 1990s which display such a transformation but with traces on every page of the path to postmodernism which they have trodden. In cultural studies, and especially literary theory, the notion of 'transgression' has been developed with regard to studies of sexuality, gender and crime. Jonathan Dollimore (1991) in the course of developing a 'cultural materialism' reviews in particular gay men's transgression and literary representation from Oscar Wilde to the present. More generally in the course of re-reading, and re-presenting, a whole range of critical theory including literary theory, Chris Stanley (1993a) has developed the notion of a whole economy of transgression including rioting, ram raiding, wrecking, joyriding, raving and other mainly male youth cultural and delinquent practices.

Moreover, the idea of 'ban' has been central to deviancy studies since the 1950s, particularly in labelling, or symbolic interactionist, theory in the United States. We began this book by noting the continuing importance of Howard Becker's notion of 'outsiders' first developed in the collection of essays of the same name, especially his outstanding ethnography of jazz musicians and marijuana use. Theorists who followed in Becker's footsteps such as David Matza (1969) have subsequently, in the process of their own descriptions of 'delinquency and drift' of youth, and especially the process of their 'becoming deviant', elaborated the notion of the 'ban'. As new and critical criminologists Ian Taylor, Paul Walton and Jock Young (1973) noted, 'ban alters the nature of the activity being engaged in: it is the force of the state criminalising an activity, proscribing it specifically as beyond the bound of law'. Colin Summer, initially reading law and deviance from an Althusserian Marxist perspective (1979), developed the concept of 'censure' (1990) in order to try to overcome the problems of critical criminology's self-confessed[12] overreaction to earlier perceived failures of new deviancy theory's attempts to theorise and analyse the quasi-legal force of the 'label' and the 'ban'. This 'force of law' has been focused on by less criminologically founded sociological and literary theorists of culture, too. As it is stated in the trans- lator's introduction to Pierre Bourdieu's attempts (1987) to pro- vide a sociology of the juridical, 'the "force of law", the quasi- magnetic pull of the legal field' is a mysterious phenomenon; for Bourdieu there are in operation 'deep structures within the juridical field', what Bourdieu terms 'habitus', which are specific to the legal field and cannot be understood as simple reflections of relations in other social realms. Deconstructionist Jacques Derrida has taken up this concept of 'the force of law' (1992), as well as others who have closely followed him in jurisprudence,[13] though with, pre- dictably, far greater focus on the play of language than Bourdieu.

Foucault's notion of 'discipline' is, of course, critical here, too, with its implications for more Marxist-based, and sociologically based, theories of power, as we saw in section 1. However, the term 'field' itself is also illuminated by Foucault's notion of discursive field[14] and the disciplining, regulating and policing of various domains.[15] One disciplinary 'field' where the 'force of law' has been studied, particularly in the United States, is in 'com- munication and law'. Separately, communication and media studies researchers have studied law or legal issues, and legal scholars studied communication, or media, law, but the two have also come together.[16] Topics such as libel and rights to freedom of expression have been studied within frameworks which make use of communication research and methodologies. This type of approach identifies 'communication and law' as an emerging field in itself, a sort of sub-genre of the sociology of law which we discussed in section 2 as itself giving birth to 'law and popular culture'. As applied research it focuses on areas like the influence of pornographic communications and the prejudicial effect of pre- trial publicity in the media, and its aim is often to question com- munication assumptions which appear to be inherent in law. This is

not, though, an exploration in theoretical terms of the transformation of society by electronic technologies, particularly television and cinema, as has been pursued elsewhere by writers like
Ethan Katsch (1989) and which we will explore in more depth in
section 4. Instead, the field of 'communications and law' represents another flawed attempt to come to terms with a new non-
traditional 'black letter law' and crime 'field', but without radically
changing the positivist approach which dominates in legal
formalism and conventional criminology.

A sphere where this problem has been to some extent overcome
and where law, crime and deviance have received a more accurate,
contextual form of study is in criminal law itself, and especially the
sociological and historical study of law and crime 'in context'.[17]
Alan Hunt has argued this in his study of the sociology of
governance, as we have already noted, in the exploration he has
undertaken of the law and 'consuming passions' where he sees the
regulation of conspicuous consumption provided, formally, by
sumptuary laws in medieval and early modern Europe continuing,
in other forms, into the twentieth century. Hunt suggests that we
are unable to understand legal and social regulation in 'the present'
without a comprehensive rigorous and detailed Foucauldian
'archeology' of the past. The crucial historical break, or watershed,
for the birth of 'law and popular culture', though, may not be quite
so far back into the 'murky waters' of the founding conditions of
modernity in history. Instead if we cast our net back to the end of
the nineteenth century, at least in the case of the advanced industrialised nations such as Britain and the other European states, to
the period 1880–1920, we can begin to unravel the 'birth of popular culture' in its modern, or even postmodern, form.

I want to pursue these questions in various areas during the
remainder of this book, but for the time being let us make reference
to the case study of the birth of association football. In the general
context of the social history of licensing and legal and social regulation of pleasure and leisure, Geoffrey Pearson (1978a, 1978b and
1983) has shown how the traditional mass entertainment of
working-class men and boys became a central focus for moral
panic, or what he terms 'respectable fears', and also how riot and
questions of control of the 'crowd' or the 'mob' have been central
features of modernity since at least the beginnings of the British
industrial revolution in the 1750s. In the context of more wide-

spread study of sporting regulation in social history, Wray Vamplew (1979, 1980 and 1988) has shown how crowd behaviour was regulated at British professional soccer matches and cricket games before World War I. Hugh Cunningham[18] has further commented, in the course of a review of Tony Mason's history of professional soccer in England from 1863 to 1915 (Cunningham 1980) that control over soccer players in the nineteenth century could be legitimised by reference to the football clubs' moral function of overseeing the problem of working-class youth. I would want to argue that soccer culture was clearly a 'site of intervention' in the nineteenth century when regulation of leisure increasingly meant the 'disciplining' and 'policing' – in a Foucauldian sense – of working-class culture in such a way that, as indeed Foucault himself argues,[19] 'respectable' and 'rough' (or 'criminal' or 'dangerous') became dividing, and divisive, categories for control of the labouring population.

Throughout the nineteenth century popular leisure practices were 'sites of intervention' for the making of the professionalised police force as Robert Storch has conclusively shown (1981) and, as Gareth Stedman Jones has demonstrated in many studies, for the remaking of the English working class (1974 and 1984). The remaking of 'popular' leisure, and more specifically, 'working class' leisure, in the 1880s and 1890s in Britain involved the birth of many elements of what constitutes the 'Popular Culture Industry' at the end of the twentieth century. A commercialised and commodified popular culture did not, of course, emerge without long, historical resistance.[20] Specifically in the area of football culture there was considerable – if uneven – development of 'consensus' in working-class culture,[21] in the same way as, by the end of the nineteenth century, there was the parallel development of the music hall in working-class culture[22] and, eventually, by the beginning of the twentieth century, a grudging acceptance of the 'discipline' of a professionalised police force.[23]

We began this section by considering the somewhat unlikely collision, and separation of deviancy theory and cultural studies, and we have traced some of the branches of the subsequent academic and cultural history. In the 1970s at the Centre for Contemporary Cultural Studies at the University of Birmingham, Chas Critcher, John Clarke and others (Critcher 1979) took up, theoretically and empirically, precisely these issues with regard to the historical regulation of leisure and 'social control' in the nineteenth-century working-class cultures around soccer. At the same time they were, with other colleagues in the National Deviancy Conference such as Geoff Mungham and Geoffrey Pearson (Mungham and Pearson 1976), extending such histories into the 'present', focusing on such phenomena as racism in British working-class culture ('Paki-bashing') and sexism in 1970s British working-class club and disco culture, as well as contemporary football hooliganism. In the next section, the first of Part II of the book, I want to bring these debates into a sharper theoretical and empirical focus, especially around the terms popular and unpopular, rather than just legal and illegal, straight and deviant. I want to pose the question: it may be popular but is it legal? and, also, ask: it may be legal but is it art?

Notes

1 See Redhead 1993c.
2 As an alternative title to 'law and popular culture' the field itself, and the work of the Manchester Institute for Popular Culture in particular, has acquired the label 'cultural criminology' in the writings of academic commentators such as Chris Stanley who are also working in this space between various established disciplines. Stanley, capturing the feel of these border/bad lands, has called his own, yet to be published, volume based on his Ph.D. thesis at the University of Kent, *Outwith the Law*.
3 See Hall *et al.* 1978. On the continuing, and sometimes forgotten, importance of this book and its analysis of 'folk devils', 'moral panics' and 'law and order campaigns' in the 1990s see McRobbie 1994.
4 See, for a summary of the ramifications of this debate, my own chapter on realism, idealism, criminal justice debates and 'law and deviance' in Brown 1984.
5 See Gilroy 1993a and 1993b.
6 Frith fought a long battle within the University of Warwick Department of Sociology over the status of his writings on popular music (whether they should be included in debates over academic tenure, and so on). Ironically, years later when he became Director of the John Logie Baird Centre at the University of Strathclyde he was appointed Professor of English not Sociology.
7 Corrigan's sheer physical size made the notion of his conducting an ethnographic study of Sunderland's teenagers somewhat mind-

boggling. In turn, when Corrigan told his 'subjects' that he was writing a book on them, they were astonished, telling the author that only 'dead' people wrote books!

8 See Fine *et al.* 1979.

9 See Robins and Cohen 1978.

10 See the excellent books which Dave Robins went on to write as a single author, branching out from the joint ethnographic work on football, culture, crime and youth with Cohen (Robins 1984 and 1992).

11 Centres such as Alan Tomlinson's Chelsea School Research Centre at what is now the University of Brighton and Richard Gruneau's 1980s Centre For Sport and Leisure Studies at Queen's University in Kingston, Canada, emulated many of the features of the earlier CCCS. Both Tomlinson and Gruneau, personally and professionally, helped to make leisure studies an important and rigorous academic forum on an international scale as well as forging theoretical paths through more established disciplines such as sociology of sport and associated professional organisations, in particular the North American Society for the Sociology of Sport. A former CCCS participant, Tony Bennett, has done something similar for the study of cultural policy through the Institute for Cultural Policy Studies at Griffith University in Brisbane, Australia.

12 See for instance Ian Taylor's book based on his Ph.D. thesis at Sheffield University (1981). Taylor himself led much of the political and theoretical work in the 1980s and 1990s which moved out of deviancy studies and towards an engagement with, and relocation of, post-CCCS cultural studies, some of it in conjunction with the Centre for Research on Culture and Society while he was a Professor at Carleton University in Ottawa, Canada. In truth, though, Taylor's writing was always as 'culturally' oriented as 'criminological'; see, as an instance, his essay, co-written with Dave Wall, on popular and youth culture 'beyond the skinheads' which focused on 'glamrock', in Mungham and Pearson 1976.

13 See Cornell 1992

14 See my essay 'Policing the field' (Redhead 1986) for linguistic play on this Foucauldian idea and, in the context of public and private spheres of society and the juridification of sport, on the more literal meaning of 'field' as in 'playing field'.

15 See my application of this in *The End-of-the-Century Party* (Redhead 1990) to the post-1950s social domain of popular music culture much as Foucault, and cultural theorists like Frank Mort (1980) who have followed in these tracks, have done in analysing and historically excavating the 'domain of the sexual'.

16 See Cohen and Gleason 1990.

17 See Norrie 1993, and Wells *et al.* 1992, for 'law in context' textbooks

which are exemplary in this regard.

18 See also Cunningham 1982.

19 See my essay on discourses of soccer hooliganism (Redhead 1991) for
    an application to the history of studies of football disorder of
    Foucault's idea in his 'birth of the prison' study that the modern
    penitentiary helped to construct socially a divide between the 'rough'
    and the 'respectable' working class.

20 See Malcolmson 1982, and Yeo and Yeo 1981.

21 See Korr 1978, Baker 1979, and Moorhouse 1984.

22 See the work of Peter Bailey and others (1986).

23 See Kettle and Hodges 1981, for a very readable historical introduction
    and a skilful demonstration of the fracturing of this 'consensus' in the
    early 1980s riots on the British mainland.

Part II

The disappearance of law into popular culture

In the post-literate millenium technology finally will sweep away all resistance to meaning and all constraints beyond the individual.

David Lange, Professor of Law, Duke University, USA (Lange, 1992)

4

The law of art and the art of the law

In section 2 the connections in the history of academic work in jurisprudence and sociology of law on 'law and popular culture' between popular music, film, the arts and the law and lawyers were explored. In section 3 the connections in the history of academic work on sociology of deviance and cultural studies on 'law and popular culture' were revealed to contain a mutual concern with transgression and rule, or law, breaking and the societal reaction to such perceived transgression. It has been emphasised already in this book that work in the reconstitution of these various fields, the terrain which I have designated as 'law and popular culture', has tended to be limited by trying to develop a 'legal theory of popular culture' rather than an aesthetics of law, or what might be called the (pop) art of law. In section 5 it will be argued that, though the search for an aesthetics of law is a necessary condition of progress in this area, it is not yet a sufficient condition. An 'erotics of law', a popular cultural theory of legal desire, as section 5 makes clear, is also required. Before embarking in more detail on this task in other sections of Part II, in this section I want to consider some concrete areas of both civil and criminal law where law and popular culture collide and where issues of law, deviance and transgression are produced and where what Howard Becker and other labelling theorists termed moral entrepreneurship is organised through diverse 'respectable fears', 'moral panics' and 'law and order campaigns'.

Let us first begin this enterprise of building the blocks of a theory of an art of law by looking at the 'law of art'. Taking one British instance on what can be called 'the state and the art' is instructive. It seems like a political lifetime since 1982 when Richard Hoggart, author of *The Uses of Literacy* in the 1950s and the first Director (that is, before Stuart Hall) of the Centre for Contemporary

Cultural Studies at the University of Birmingham in the 1960s, was removed as Head of the Arts Council of Britain by Margaret Thatcher's first government, to be replaced by the former editor of *The Times*, William Rees-Mogg. Hoggart, along with Raymond Williams's theoretical and literary texts, provided Stuart Hall and the CCCS with a rich legacy from the New Left of the postwar political spectrum to exploit, within and without the academy, in the study of culture and cultural politics. However, after many years of free market policies and what Hall theorised as 'the great moving right show', by the early 1990s it was the New Right which had increasingly managed to elevate 'culture' to the centre of the political stage.

It was, for instance, David Mellor who, having played the role of Arts Minister in the third Thatcher government in the United Kingdom, eventually succeeded to the position of Britain's first 'Minister of Culture'. One result of the British General Election of April 1992 was the creation of a new ministry, named the Department of National Heritage, covering the areas of parks, heritage, tourism, the 'arts', film, sport and broadcasting. David Mellor's initial pronouncement as Secretary of State was to the effect that alongside the serious enterprise of wealth creation in the new conservative – or Conservative – era of the 1990s there should be room for some measure of pleasure and culture – in short, 'fun'. Mellor's short stay in office was ended when tabloid newspaper revelations about his private life – an affair with an actress which was exposed by tapped telephone calls – forced his eventual resignation after nightly television news items featured ever more lurid details over several months, although he managed to turn the tables by making a lucrative career out of broadcasting[1] and journalism once his ministerial career ended. Mellor, in fact, helped to pilot the Broadcasting Bill – what became the Broadcasting Act, 1990 – through when he was at the Home Office and was credited at the time with winning the Independent Television Commission a degree of discretion to award the ITV franchises for quality as well as money, and with eschewing the more widespread Conservative Party rhetoric against public service broadcasting. The issue of broadcasting, especially the renewal of the British Broadcasting Corporation charter in 1995, was prominent in critical discussion of whether this particular government minister could remain at his post considering the sensitivity of the issue of the regulation of the press and broadcasting and the ever-widening debate about privacy on the one hand and the right to publish, to freedom of information and to freedom of speech on the other. The question of whether self-regulation – in the form of the Press Complaints Commission – should continue is central. The issues surrounding law and the media have, manifestly, never been so intertwined.

Indeed, the question of law and (media) representation is a pervasive one. It alludes both to the idea of the relationship of law – national and international legal systems – to the means of communication or the mass media in general, and also, to crucially, the jurisprudential question of 'what is the object of law?': that is, what does law represent? It has been widely assumed in contemporary critical legal and social theory that there is increasingly a crisis in representation; indeed some of the debates about postmodernity and modernity, and modernism and postmodernism, have centred on such issues as the defining characteristic of the end-of-the-millennium mood in which we are now supposed to find ourselves.

In this context Jane Gaines's book *Contested Culture* (1992), originally entitled *Likeness and the Law: Image Properties in the Industrial Age* when it was first published in the United States in 1991, is an important contribution to legal and cultural studies in this area, and the British Film Institute should be congratulated for bringing its theoretical focus on 1970s British cultural studies back home. Nevertheless, part of the difficulty with assessing the overall potential influence of the relatively few existing books on 'law and popular culture', as we saw in section 2, is that the general theoretical traditions within which they are framed have been subjected to sustained critique in other fields. In Gaines's case the author's determination to 'demonstrate the effectivity of the Marxist features of British cultural studies' rings rather hollow in the early 1990s. Her study proceeds through case studies of the images surrounding Oscar Wilde, Jackie Onassis, Nancy Sinatra, Superman and Dracula but leaves the reader in something of a theoretical time warp. This feeling of participating in conversations from an earlier era – pre-1980s – is further emphasised in Gaines's concentration on American legal theory in the form of the Conference on Critical Legal Studies (CCLS) where, like the European Critical Legal Studies group (ECCLS), European Marxism has retained its hegemony rather than the later British variety of Critical Legal Studies (CLS) where debates around psychoanalysis, literary

theory, postmodernism and a range of feminisms have pushed
critical cultural analysis in some very different directions. Despite
such major misgivings, there are rich and thoroughgoing investiga-
tions in the 'empirical' chapters into the legal fields of privacy,
trademark protection, 'star' contracts and so on, which should be
compulsory reading for anyone interested in intellectual property
law and the communications media, despite any theoretical limita-
tions. Further, if it is read together with Celia Lury's treatise on
'cultural rights' (Lury 1993) – which argues, importantly, that the
regime of cultural rights has shifted from one ordered by
authorship, originality and copyright to one which is defined by
branding, simulation and trademark – there is much specific coverage of, and stimulating questions for, an emerging field of socio-
legal study of law and 'culture-as-property'. If we take the 'black
letter law' field of intellectual property, in particular copyright and
its subdivisions like music and video private copying,[2] and consider
it in the context of a certain area of enquiry such as cultural
industries, the two-sided nature of the question of law and representation can be seen.

The thrust of legal change in this domain has come from a
delayed reaction to the new communication technologies, for
instance, video, computers, cable and satellite television. The two
statutes in Britain which mark the historical period in question (the
post-1945 era) are the Copyright Act, 1956 and the Copyright,
Designs and Patents Act, 1988. The most recent problem in the lag
between technological change and legal discourse is that of
'sampling' in the music industry, where sections of other material
are included in, or even constitute, new commodities for sale. The
questions about 'rights' discussed in the copyright debates are
varied and mixed: they relate firstly, to 'interests' such as those of
the author, the publisher and the public and, secondly, to categories
of rights such as 'moral rights', 'rental rights' and copyright.

Several traditions in specific legal systems can be recognised. The
French law is different from that of England, and the United States
– which Jane Gaines concentrates on – is separate again, and,
though greater convergence is now found as a result of the Copyright, Designs and Patents Act, 1988 introducing 'moral rights' into
English law for the first time, there are significant differences of
approach. A continuing debate in the late 1970s and early 1980s
between Bernard Edelman (France) and Paul Q. Hirst (England)

focused on the perceived differences and highlighted problems in the project for a general (Marxist) theory of law in jurisprudence as Paul Hirst and the translator of Edelman's text, Elizabeth Kingdom,[3] foresaw. It is this debate about ownership of the image which pervades the whole of Jane Gaines's book, and indeed constitutes a project which has threatened for some years now to unleash itself into the arena of struggle which is 'law and popular culture'. In an era of copying and plagiarism as art – following pop artist Andy Warhol's multiple reproductions of the images of Marilyn Monroe and Campbell's soup tins, amongst others, in the 1960s – the question of ownership of the image remains critical.

Bernard Edelman, a French barrister influenced by the structuralist–Marxist theories of Louis Althusser, wrote his own idiosyncratic account in one of the first major works in the field of 'law and popular culture', *Ownership of the Image: Elements for a Marxist Theory of Law* (Edelman, 1979), first published in French in 1973. As Vincent Porter (1979/80) pointed out at the time in *Screen*, the film theory journal where many of the debates about 'law and popular culture' have taken place, Edelman's book raises a number of problems for the Anglo-American reader in terms of both the difficulties, if not the impossibility, of constructing a general theory of law and the specific relevance of such a theory of law to a copyright legal tradition as exhibited in France. The problems, as Porter noted, are therefore general and particular. Edelman's view of copyright law is also coloured by his choice of technological focus, which is that of photography as signified in the French title of the book – *Le Droit saisi par la photographie* – a pun on the law being seized, or caught, by photography; that is law being surprised or caught out by technology. It is worth noting that in France the law was indeed caught out by the advent of photography and the need to ensure adequate protection not only for the photographic trade but for the cinema in the form of the international film business. The law of photography is Edelman's chosen area because it fits into the heart of his interests as a theorist of jurisprudence, and as a legal advocate, in that it concerns the law of property and, furthermore, the definition of the status and rights of labour. It is this law which is caught out by the pre-existing definitions of a property right in representations of things. Edelman's focus on photography has often been taken up in other works. For instance, John Tagg, in two seminal articles on power, photography and the law (1980 and 1980/81) made use of Edelman's *Ownership of the Image* while constructing a Foucauldian history of the development of photography as a means of surveillance and record. Jane Gaines makes fascinating use of Edelman in each chapter of her book *Contested Culture*, which focuses not on photography as such but on the relationship between, as the subtitle suggests, 'the image, the voice and the law'.

Bernard Edelman's own thesis on the law being caught out by photography serves as a symbolic warning, as David Saunders (1988, 1992) has argued, about a threat to other areas such as literary rights, since the whole domain of intellectual property is becoming transformed by audio-visual production and information technology. Edelman analyses the way in which French law came to recognise photography not as the working of a mechanical apparatus but as an act of individual human creation which when it produced an original work could be protected on behalf of the author under the system of 'moral rights' enshrined in the French legal system. This is not merely a jurisprudential problem about conventional property rights and legal relations but a more critical and significant historical question which photography posed for French law about literary and artistic property. Photography caught out the philosophy of law in France. In that country literary and painterly property is established, in law, by the category of 'author's right'. English copyright law (going back further than merely the Copyright Act, 1956 to the Copyright Act, 1911 and even to legislation at the beginning of the eighteenth century) differed from this – prior to the Copyright Designs and Patents Act, 1988 and even afterwards to a large extent – in that it did not derive the right to control the reproduction of a work from the idea that it is the product of a creative subject; nor for that matter did it have to recognise such a right in the subject who actually makes the work.

The theory of the legal subject – indeed subject positions – is a complex field. It forms the basis of much of the dispute between Bernard Edelman and Paul Hirst and receives an illuminating discussion in Hirst's introduction to the English edition of Edelman's book especially in as much as it related to Marxist jurisprudential debates in the 1970s which Hirst himself did so much to promote in his book *On Law and Ideology*. Jane Gaines uses this aspect of the argument as a constant backdrop to her own intuitions on law, media and popular culture. Paul Hirst's important contention is that, however interesting, though flawed, the Edelman analysis of the law relating to photography and cinema may be in demonstrating the role of the subject-in-law, Edelman is limited by confining his endeavours to French law – the case study would not, Hirst claims, work so well, if indeed at all, in the context of English law. The distinction between copyright in English law and author's right in French law is, for Hirst, all-important. In English law, copyright, essentially, has been specified as certain definite capacities of legal subjects who meet certain conditions, and the basis of such a right is not morality but statute. Thus the 'author' of a photograph who had the copyright was the owner of the material on which the photograph was made at the time it was taken (author here of course need not then be creator). In English law, copyright as a right based on creativity – a creative subject-in-law – does not exist, unlike 'moral rights' traditions where the creative, artistic subject is a paramount concept.

The moral rights tradition focusing on individual not multiple authors in French law is based on the principle of a right of personality which in that legal system recognises the property relation of work to author. It is conceived in continental jurisprudence in Europe as an inalienable right. In the change in English law under the Copyright, Designs and Patents Act, 1988 moral rights (the right of the author to be identified as the author of the work – which must in any case be asserted – and the right to object to derogatory treatment of the work) are recognised for the first time. But United Kingdom authors can be required to sign away their moral rights by contract. The Copyright, Designs and Patents Act, 1988 also further undermines the import of the idea of author's right into English traditional copyright discourse by creating rental rights for sound recordings, films and computer programmes which are rights for producers/organisations not individual, physical authors. It has been concluded in at least one analysis of the emergence of the 1988 Act, which came into force in 1989, that the rhetoric of the legislation had indeed been to promote the concept of author's right yet the Act had in practice been a compromise between certain social interests – the author, the publisher and the user – and that far from extending the rights of authors the Act deprives them of many rights to which they have traditionally been considered to have been entitled. For Vincent Porter (1989) the Act was in fact a clear case of the 'triumph of expediency over principle'.

In the music industry the question of changes in technologies and changes in culturally derived legal relations of copyright is deeply complex and only recently has there been much of a literature which analyses it seriously.[4] In Jane Gaines's text it is the film industry rather than the popular music business which is theorised but the parallels and comparisons are instructive. The Copyright, Designs and Patents Act, 1988 in principle gives the creators of music certain rights, including the right to control the public performance and broadcasting of their works. This means that every time a song is performed in public, broadcast or included in a cable programme service, the songwriter has to have given his or her permission for its use and to have negotiated an appropriate royalty. In practice, the popular music industry throughout the world is rife with 'sampling' of other works, few of which are compensated, not to mention a variety of other practices which undermine author's right. The record industry failed, after years of campaigning, to get a blank tape levy into the Copyright, Designs and Patents Bill as it was passing through Parliament. What is certain is that, as the music and cultural critic Simon Frith (1987 and 1993) has argued on a number of occasions, in the case of the music industry the age of manufacture is at an end since businesses now depend for their survival on creating rights (and then exploiting them) rather than 'things' (records, T-shirts, videos) for sale to customers. Legal disputes, though, continue in the music industry, for example the conflict involving the Mechanical Copyright Protection Society over the level of royalties due to publishers for the right to record their songs which was taken to the Copyright Tribunal. Debate has been long and litigious, too, about questions such as, in Avron Levine White's words (1987), 'Who owns the song?'

Property ownership rights have become considerably more complex as technologies develop apace and cultural theories collapse the binary distinctions of fake/authentic, and original/ copy. The threat to those rights of intellectual property constituted by sampling (and many other practices such as home taping) is also sometimes discussed in terms of art/plagiarism – in other words either as straightforward 'theft' or as part of the thinking processes of a creative subject. Authorship, creative subjectivity and 'cultural appropriation'[5] are increasingly problematic concepts in popular culture – and especially popular music – today now that technology in the age of digital reproduction has become more accessible, and relatively cheap, and the process of globalisation is so pervasive. The legal subject of the creative process, consequently, is more and more difficult to identify, especially as the role of musician/DJ/ producer/engineer becomes ever more blurred.[6] Law courts have resounded to the clash of litigants' arguments over who authored what for years. For instance, Milli Vanilli were sued for not being 'real' musicians: that is, because they did not play on their own records. Musicians like George Clinton and James Brown could, potentially, be involved in thousands of court cases against people who have sampled them over and above the cases which have already occurred. However, it has become clear to some who have been accused of using unauthorised samples, like the Shut Up and Dance record label responsible for the compilation LP *Fuck Off and Die* in February 1992 and the chart single 'Raving I'm raving', that 'the rules and regulations concerning sample clearance are rigidly set out to the detriment of small labels'. In contemporary dance music in particular small dance labels have been the most significant promoters of sampling; in 1990 De Construction records, for instance, had massive commercial success with an 'Italian house' single, Black Box's 'Ride on Time', which featured a sample from American disco singer Loleatta Holloway, causing considerable legal and musical dispute.[7]

In Jane Gaines's book 'high' film theory dominates rather than the 'low' theory of popular music and cultural criticism. This break between 'elite' and 'mass' – and its reconstitution in an age of what some commentators have seen as a 'postmodern' flattening of the bridge across such a great divide – itself constitutes part of the problem of ownership of the image which Gaines seeks to study. Consequently, her work is perhaps more useful in application to some aspects of the media and cultural industries than others. Nevertheless, it stands as a gallant, if flawed, enterprise to broaden and theoretically deepen the study of law and (popular) culture. But, as is clear from her book, the debates about popular music, plunder and 'theft' in popular culture, and the concepts of legality and illegality, have taken place, moreover, against the backcloth of controversy in the social and human sciences generally. Within, and beyond, such disciplinary boundaries argument about the concepts of postmodernism, postmodernity and postmodernisation has, as we shall see in section 5, included 'postmodern jurisprudence', postmodern law or postmodern legal theory. Two facets of such a field are apparent in the analysis of copyright in postmodern jurisprudence made by Costas Douzinas and Ronnie Warrington, with Shaun McVeigh (1991). Firstly, there is the jokiness of the tone and the disparate, fragmented arrangement of the various parts of the texts. Secondly, there is the importation of literary theory in general, and in particular the theories and practices of deconstruction associated with writers such as Jacques Derrida, to the analysis of property rights – in this case, appropriately enough, literary rights and copyright. The title of the chapter,[8] which was earlier in its literary life a conference paper performed (or 'read') 'live' collectively by all three authors dressed head to toe in black, is 'Suspended sentences', a phrase especially pregnant with double meaning, in linguistic and legal, literary and jurisprudential senses. Legal and other texts abound in the chapter, including the positivist, 'black letter' law of copyright and mock letters to and from academic legal journals as well as discourses on and from a literary work, Herman Melville's *Billy Budd*, in a fascinating play on 'texts of the law and law of texts'. As we shall see in section 5 it is more accurate to describe Warrington and Douzinas's jurisprudential enterprise as rooted in post-structuralism rather than in postmodernism, but in any case it emphasises the extent to which legal theory has been invaded by cultural studies and debates about the 'postmodern'.

There is clearly, too, not just in the fields of popular music and film but across a wide popular cultural spectrum, the question of 'electronic piracy' as John Chesterman and Andy Lipman put it in their book on the extremely extensive 'DIY crime of the century' (1988). Old ideas of copyright have, as we have already seen, been blown into smithereens by rapid changes in new technologies. But electronic guerrillas are burgeoning as a result and societal and legal reactions have been many and divergent, at least in Western industrialised countries. In England and Wales for example, the Computer Misuse Act, 1990 was passed after a Private Member's Bill introduced by Conservative MP Michael Colvin received wide support following the withdrawal of an earlier bill by another Conservative MP, Emma Nicholson. Computer hacking, phone phreaking and other deviant activities of the electronic counter-culture are discussed by Owen Bowcott and Sally Hamilton in their story of the electronic underworld (1993) and Katie Hafner and John Markoff even take the term 'cyberpunk' as the title of their book on the computer underground (1993) which they argue is the real-life version of cyberpunk. Computer hacking is perhaps the best-known example of electronic crime but such illegalities go much wider.[9] Cyberpunk writer Bruce Sterling (1993), collaborator as we noted in section 1 with William Gibson, documents contemporary cyberspace and the efforts being made by law enforcement agencies such as the Federal Bureau of Investigation and the secret services to deal with its own 'criminal' elements. Sterling shows how in the United States in one incident involving a computer games company an entirely innocent organisation could be raided, and have its computers confiscated, without the secret service arresting anyone, or accusing anyone of criminality, or even suggesting what the nature of the illegality committed might be. For Sterling this is part of the law-and-order 'crackdown' intended to serve notice on America's cyberspace villains. It was to some extent a reaction against bulletin boards, which, unlike printed publications, are not protected along USA Freedom of Information Act lines. One outcome of all this was the creation of a pressure group body, the Electronic Freedom Foundation. Bruce Sterling recounts the crackdown on hackers by concentrating mainly on tele-communication issues, such as phone fraud and 'phreaking', but his book contains virtually nothing about computer hacking as such. The focus of the book is 'cyberspace', the region in which electronic communication takes place. As Sterling (1993) puts it:

> cyberspace is the 'place' where a telephone conversation appears to occur. Not inside your actual phone, the plastic device on your desk. Not inside the other person's phone, in some other city. The *place between* the phones. The indefinite place *out there*, where the two of you, two human beings, actually meet and comunicate.

This cyberspace has been the focus for much debate and theoretical discourse about safe sex in a post-AIDS era, spawning novels[10] and short stories[11] consisting wholly of simulated examples, as well as documentary texts, and, through them, assertions of evidence of the emergence of a new age of what Mark Poster (1990) has theorised as 'the mode of information'.[12] The regulatory regimes which are necessary to create surveillance of such space can be envisaged as 'cyber law', the policing of cyberspace and the dawn of the disappearance of the 'body-in-law'. One moral panic which has emerged in this area is over the fusion of eroticism and computer virtual reality.[13] Technological developments which have made 'cyber-sex' a possibility rather than a science fiction myth have led to moralistic campaigns (though this is safe sex taken to its ultimate). As police struggle to control pornography on computer bulletin boards, and new legislation is campaigned for by pressure groups who perceive there to be a gap in the law, virtual reality sex through interactive sex-games and 'teledildonic suits' is threatening to make legal and social regulation impossible, a state of 'virtual law'. In Britain, until recently, much of the conflict that has arisen has involved 'law-and-order campaigns' over censorship of young people's consumption of telephone sex services, which are now common throughout Western culture. In France, Minitel video keyboards attached to telephone sytems in the 1980s transformed the notion of safe sex services, creating as Guy Sitbon has argued (1988) 'another country' in the age of disease prevention where a life that is not 'telematic' is not worth living. Leslie Dick (1989) in her short story based on simulated Minitel conversations, neatly summarises the significance of the system in her preface:

> It was a love affair by electronics, a love affair by *Minitel*, the Paris computer system that offers interactive communication between users, who type texts onto their computer screens, these texts transmitted instantly from monitor to monitor, through the telephone lines of the system . . . The *Minitel* system is low tech: you can book theatre tickets, or find out railway timetables, you can check your bank balance and then, you can give in, accept the invitation posed by incessant, all-pervasive advertising, and dial 3615, the number that lets you into the network of lovers, those who pursue sexual gratification through the exchange of texts, only. No kiss, no touch; no object; no letter, or scrap of silk; no image, no voice, even, no trace of the body. The body vanishes, leaving a flickering screen, rows of words inscribed in fugitive white light.

Bruce Sterling's story of the regulation of cyberspace begins with a history of the telephone and shows how the failure of the AT & T phone system in the United States in 1990 – not an incident in fact caused by unauthorised hackers, although technically it could have been – became the event which prompted the US security authorities to teach the 'punks' of the digital underworld a lesson. The crackdown included surveillance of computer bulletin boards, wire tapping and other tricks prior to high-profile court cases and, as Sterling shows, a more regulated environment for the global hackers, phreakers and other electronic outlaws of the 1990s.

'Law and order campaigns' about other screens – as well as computer – such as film, video and television abound. In Britain in 1992 the BBC was criticised by the Broadcasting Standards Council (chaired by Lord Rees-Mogg) for showing a film adaptation of *A Time to Dance*, one of the novels written by broadcaster and president of the National Campaign for the Arts, Melvyn Bragg, which contained a rape scene (as well as other incidents of 'sexual explicitness', complaints about which were not upheld) on the grounds that children were still likely to be watching even though the broadcasts were after the 'watershed' time of 9 p.m. 'Explicitness' is often tolerated by the council on the basis that warnings have been broadcast prior to programmes, though the BBC was taken to task over scenes from a television version of Ann Oakley's book *The Men's Room* in 1992.

De-regulation of the media and rapid developments in satellite technology have made national legal regulation ever more problematic. For example, in 1993 the 'hard core' satellite pornography channel Red Hot Dutch, broadcasting from Denmark, was the site of a struggle over whether a national government (that is, in this case the United Kingdom) can regulate a channel which can be received in any European country. The channel could be seen, courtesy of Continental Television, in 22,000 homes in Britain and as far as the company was concerned 'operated under the sensible pornography and broadcasting laws of Denmark' which the majority of its directors thought should be good enough for Britain under European Union law. The Broadcasting Standards Council issued a critical report about the films broadcast on the channel and called on the United Kingdom government to ban the station. Legal action in the English courts was taken against the company. However, under European Union regulations, Red Hot Dutch is legal all over Europe if it is acceptable in only one country, and a government taking unilateral action risks being accused of reneging on European Union broadcasting agreements. A ban ordered by Peter Brooke, National Heritage Secretary of State, on the subscription service in Britain did not prevent the estimated tens of thousands of subscribers from receiving the station, but the supply of decoders and programme material, and advertising, was made illegal. The ban was upheld by the High Court and, subsequently, the Court of Appeal, though the case was then referred to the European Court of Justice in Luxembourg. In fact in July 1993 the Red Hot Dutch satellite station ceased broadcasting for a while after Danish authorities withdrew the satellite link following non-payment of bills. Even when broadcasting resumed – with the help of 'smart cards' openly advertised in the continental European press – citizens in the United Kingdom were still able to receive signals of these, and for that matter other, international satellite programmes.

Censorship, or regulation or de-regulation, debates are of course nothing new. According to feminist criminologist Beverley Brown (1982) the British Board of Film Censors has long constituted a 'curious arrangement', and changing technologies and the difficulties of enforcing laws in this field make it no less curious today. The pre-satellite and pre-video 1979 Report of the Committee on Obscenity and Film Censorship (known as the Williams Committee after its chairman, Bernard Williams) was based on liberal philosophy, inherited from nineteenth-century theorists such as Jeremy Bentham and – in particular – John Stuart Mill, which held that conduct should not be suppressed by law unless it can be convincingly demonstrated to carry a real prospect of harm to others, especially children. This official report,[14] never in fact legislated upon, is perhaps best seen as the last throw of the 'permissiveness'[15] movement which produced what the National Deviancy Conference in the late 1970s (NDC 1980) – building on earlier deviancy theory concepts such as 'moral legislation' or the 'legislation of morality' called the 'sixties legislation', or 'the legislation of consent'. The 'sixties legislation' began with the 1957 Report of the Committee on Homosexual Offences and Prostitution (known as the Wolfenden Report after its chairman, John Wolfenden). It is certainly the modern liberal philosophical question of the distinction between the two spheres – the public (which should be regulated) and the private (which unless the harm condition operates should not) – which absorbed much of the 1980s debate[16] first stimulated by the H. L. A. Hart (1963) argument with Patrick Devlin in the late 1950s and early 1960s following the reception of the Wolfenden Committee's recommendations.

The campaigns in Britain in the mid-1980s over the issue of 'video nasties' turned on the possible collapse of this modern distinction as the Video Recordings Act, 1984 passed into law regulating what consumers could watch on video in the privacy of their own homes. The moral panic[17] which led to this Private Member's Bill being brought in by Conservative MP, Graham Bright, was reminiscent of the campaign in the 1950s to outlaw American horror comics which Martin Barker has studied so effectively (1983, 1984a and 1984b). This campaign to expel a certain 1950s form of 'Americanisation' from the British Isles had also led to the creation of a new law, the Children and Young Persons (Harmful Publications) Act, 1955.

As much as the vexed issue of sex and the screen, the question of video, television and film violence has a long and chequered history[18] and involves the same competing social forces of libertarianism (of Right and Left), conservatism, liberalism and different feminisms. Feminist debates have raged over the issue of pornography for decades alongside competing changing positions of the moral Right ('moral majority' and 'new moralism' were transatlantic terms in the Reagan and Thatcher years of the 1980s) and liberal thought.[19] They failed to ask, and answer satisfactorily, questions such as 'What is sexuality?' 'What is desire?'[20] and 'What is the body?',[21] themes which we will explore in section 5 in the search for the foundations of an erotics of law.

Today, though, there is increasingly a polarisation between, on the one hand, a 1990s backlash against earlier feminisms which queries the attribution of the term 'rape' to many sexual encounters such as 'date rape' – as Camille Paglia (1993) and Katie Roiphe (1993) have been given such a mass media platform to proclaim – and, on the other hand, the lower profile persistence of a more orthodox position which sees pornography as 'visual rape'.[22] Feminist interest in pornography[23] has also attempted to ask whether there could be pornography for women[24] and whether porn is antithetical to feminism.[25] Such competing social forces can be seen to be fighting on terrain – the domain of the sexual which itself may be problematic, as Michel Foucault and others have shown.[26] The terrain of (mainly masculine) violence, too, is problematic. Periodic campaigns by governments of the New Right in particular have cemented a moral conservatism as a position to ascribe to any of the various political parties, for example on issues such as violence on screen being proclaimed to be the cause of actual violence despite the absence of persuasive evidence for anything other than a 'casual' link. Such moral crusades are not, however, completely successful. For some regulatory bodies, such as the Broadcasting Standards Council in Britain, television is justified in showing violent crime, fictional or real, since otherwise it would not reflect contemporary society.

Crackdown has occurred periodically, too, in the popular music industry. As Martin Cloonan (1991) has argued, in Britain 'the history of the censoring of popular music is in many respects the forgotten history of British popular music' and much the same could be said for the United States, notwithstanding the fact that in the USA citizens can formally appeal to the constitutional protection of freedom of speech in the First Amendment. American courts have had a history of serving as the arena for the struggle over popular music's allegedly 'corrupting' hold over youth. Since the mid-1980s, the Parents Music Resource Center (formed originally by senators' wives including Tipper Gore, wife of Al Gore, Democratic Vice-President) and right-wing fundamentalist Christians and groups like Focus on the Family have stepped up the heat of the campaign to censor 'rock music', with rap and various forms of metal (speed, thrash, death and heavy) frequently perceived as the main culprits. This had culminated in persuasion of the music industry in the USA voluntarily to attach to its products warning labels about sexually explicit and/or violent lyrics. One highly publicised case involved the parents of two youths who comitted suicide after supposedly receiving subliminal messages from Judas Priest's *Stained Class* album. In another incident heavy metal music was allegedly played by the perpetrators (Black Sabbath and Venom fans) of a teenage murder 'to get us in the mood' (as the killers were quoted as saying) to commit the offence.

The courts and prosecution agencies, however, have regularly intervened in the global market for popular music commodities. For instance, in 1992 US 'grunge' band Nirvana's 'Smells like teen spirit' and other tracks were classified as adult material by a Washington State censorship law passed in their home town of Seattle, the main urban centre associated with this late 1980s musical form, which also translated later to the European catwalks in an orgy of dressing down in large check shirts, jeans and ubiquitous long hair. This had the effect of making it illegal for fans under eighteen to buy their records and implicitly making it necessary for buyers to produce an identity card. The law extends

Washington State's book and film censorship to cover 'unsuitable' music, including much 'hard rock' and rap. The bill was initially sponsored by a state representative pressurised by a constituent who complained that her four-year-old child repeated offensive lyrics from a rap album.

Rap has probably been singled out for legal and social regulation more than any other form in the history of popular music. In late July 1992 Ice-T and his label withdrew the rapper's 'Cop killer' track from the *Body Count* (the eponymous LP from Ice-T's black speed metal offshoot) record after employees of his label, Time-Warner, had allegedly received death threats from the police. However, in defence of his First Amendment freedom Ice-T announced his intention to give away copies of the track at live shows; in 1989 he had debated on live US TV (the 'Oprah Winfrey' show) with Tipper Gore – who eventually resigned from PMRC in 1993 – on the issue of freedom of speech. The track had been issued first as an album track and then, in the spring, as part of the *Batman Returns* Time-Warner blockbuster film soundtrack. Texas police then called for a national boycott of all Time-Warner's commodities (including Disneyland as well as *Batman Returns* and the *Body Count* LP) and a Los Angeles council woman called for sales of the album to be stopped. In early July 1992, a few weeks after the Los Angeles riots, the (then) President of the USA, George Bush, spoke out 'against those who use film or records or television or video games to glorify killing law enforcement officers', while his Vice-President, Dan Quayle, blamed the company for making money from a record which, he alleged, legitimised killing of police officers, and Colonel Oliver North, speaking as President of the right-wing Freedom Alliance foundation, called for all US state governors to institute criminal proceedings against the record label. Canadian police, meanwhile, both tried to stop imports into Canada of *Body Count* and investigated the possibility of holding venue owners liable if a policeman was killed on the premises after the track was played. Time-Warner initially defended the rights of their artist on the grounds of the nineteenth-century Romantic ideology of artists' rights and creative subjects. Managing director, Jerry Levin, began the resistance campaign by writing to all his staff proclaiming that the company 'won't retreat in the face of threats of boycotts or political grandstanding . . . we stand for creative freedom. Whatever the medium – print, film, video, programming or music – we believe that the worth of what an artist or journalist has to say does not depend on pre-approval from a government official'.

The record label issued what amounted to a 3,000, word statement of support for Ice-T, defending his rights as an artist and proclaiming him to be a spokesman for a large section of American society. In fact, Time-Warner subsequently did discuss a change of policy on distributing music which was deemed 'inappropriate', and eventually Time-Warner dropped the artist in February 1993, much to the delight of Colonel North who celebrated the decision as a victory for the power of the common 'man', the 'average American shareholder'. The corporate debate in the immediate aftermath of *Body Count* in fact focused on Ice-T's 'KKK bitch' track, which was written about the sodomising of a Klu Klux Klan leader's daughter, and followed film megastar Charlton Heston's protest at a shareholders' meeting of the company where he read out the lyrics. Women's groups have long been vociferous in their feminist opposition to rap's frequent 'dissing' of women. For instance, the California Women's Law Centre managing director wrote in the *LA Times* as a response to the 'Cop killer' controversy that rap music was 'rape music' and slated Ice-T and fellow rappers for their 'misogynist hate-spew of cut and slash, beat and burn, destroy and discard'.[27] Furthermore, in 1993 'gangsta' rappers the Geto Boys, from Houston, Texas, were criticised by the Los Angeles chapter of the National Organisation for Women (NOW) because they included an anti-gay, anti-(abortion)-choice song 'The unseen' on an album called *Uncut Dope*. The National Organisation for Women claimed that the Geto Boys' song lyrics attack homosexuality and refer to women who have abortions as murderers. The band's leader argued that the song was designed to raise consciousness of the problem of young people who failed to practise safe sex and used abortion as a form of birth control, but the head of the Geto Boys' label, Rapalot Records, was also quoted as saying that 'as far as I'm concerned, people that specialise in murdering babies are a bunch of devils'. In 1990 the Geto Boys' tour promoters dropped the band after a show in Chicago was banned because of 'offensive material'.

In the summer of 1990, Florida's Federal Court declared Miami's 2 Live Crew's *As Nasty as They Wanna Be* album to be obscene and within hours a record shop assistant was arrested by the police for selling the LP to an eleven-year-old girl. Two members of 2 Live Crew were also arrested at one of their adults-only shows. The Republican governor of Florida, Bob Martinez, was the man responsible for organising the campaign against the band, first trying a racketeering case against them and then, when this failed, switching to prosecuting retail distribution outlets. As a result of the success of the Florida case, many more state legislatures considered making obligatory the music industry's voluntary code on label stickers (in 1990, twenty out of fifty were reported to be actively preparing a legal change). Subsequently, the powers given by the USA Child Protection and Enforcement Act, 1988, which allows the Justice Department to arrest anyone connected with selling or promoting an 'obscene' record, have meant that the Florida court case can be repeated across America (2 Live Crew's artistic response was to release a single, 'Banned in the USA'), and prosecutions have in fact multiplied. In May 1992, the band's album *Sports Weekend* was targeted in Omaha, Nebraska, where record shop owners were accused of selling 2 Live Crew tapes (allegedly obscene material) to minors; that is, to those under eighteen. The prosecutions arose after a pressure group naming itself 'Omahans For Decency' sent minors into shops deliberately to test that the restrictions were being applied. Charges were subsequently dropped when the shop owners pledged not to repeat the offence. Nevertheless, in January 1993 the US Supreme Court refused to review an earlier ruling by a lower court that the Florida sheriff who originally prosecuted the group had failed to prove that the *As Nasty as They Wanna Be* album was obscene. Effectively this amounted to the US Supreme Court proclaiming that the record was not obscene. Witnesses for the band testified that the LP was of artistic merit. Not all progressive critics took this line, however. American black studies professor, Houston A. Baker Jr (1993) argues in his discussion of this and other cases of the 'black urban beat' and 'rap and the law' that the music continues to 'provide sometimes stunning territorial confrontations between black urban expressivity and white law-and-order' though he criticises the eventual appeal court decision which went in favour of 2 Live Crew.

The banning of popular music commodities has been a British phenomenon as well an American one, as Martin Cloonan has rightly noted. Los Angeles rappers Niggers With Attitude (NWA)'s

Efil4Zaggin (Niggaz4life, backwards) LP was banned in the United Kingdom in 1991: and an estimated 25,000 copies of the cassettes, vinyl records and compact discs were seized by police under section 3 of the Obscene Publications Act, 1959. This was the first time the Act had been used to intervene against a major label – Island in this case – although both Crass and the Anti-Nowhere League (both radical 'indie' bands on small labels) had been successfully prosecuted under the legislation in the 1980s.

In the USA the album had entered the Billboard LP charts at number two – at that time, the highest new entry by a rap act – and their record label 4th and Broadway, and the parent company, Island, reported that it sold 1.5 million copies in America including almost a million in the first week of release. The US sales success can be attributed partly to a whole series of attempts at continued censoring of NWA activities which had already occurred in the USA; for instance, Ice-Cube, a former member of the band, courted much controversy when in 1992 he released a post-NWA album called *Death Certificate* which was criticised for racism and anti-semitism.

In June 1991 an irate retailer in Britain had complained about the NWA album, which is full of swearing and allusions to drugs, oral sex and violence, and police action followed. The case was referred to the Crown Prosecution Service, and retail chains like HMV, Virgin and Our Price immediately refused to stock the LP; in Ireland, after an intervention by the Minister of Justice, the LP was also withdrawn by distributors. Almost immediately a defence campaign was organised with a chorus of opposition from within and without the United Kingdom music industry. The Civil Liberties Trust denounced the police action, as did Alternative Tentacles, the independent label whose former radical band the Dead Kennedys had been the subject of massive censorship campaigns in the USA – and to a lesser extent in the UK – in the 1980s. Singer Sinead O'Connor also spoke out against the censorship, whilst criticising the band's sexism, saying that 'whether we like it or not, NWA are representative of a large number of people who share these kinds of attitudes towards women and violence. By banning the LP, or any similar material, people are just becoming blind to the realities of life'. She noted that, though she was

> a big fan of NWA before Ice-Cube left the group . . . their attitudes
> have become increasingly dangerous more recently. The way they

deal with women in their songs is pathetic. Having said that, no one should have the right to decide what the general public choose to listen to. In effect what the authorities have done is ban the record without going through any kind of legal procedure.

By November 1991 the dividing line between offensive and illegal was considered in law in a magistrate's court decision on the case which resulted in a victory for the group, allowing the stock of the LP to be restored to record shop shelves. Costs were awarded to Island, and to Polygram Records, the distributors of the record, to be paid by the Metropolitan Police who had first launched the raids in June of that year. Geoffrey Robertson, the QC who exposed the United Kingdom government's role in selling arms to Iraq prior to the Gulf War in the Matrix–Churchill affair[28] and a textbook authority on obscenity law,[29] media law[30] and civil liberties[31] defended NWA in court (as he had, as a junior to John Mortimer, helped to defend the editors of *Oz* magazine twenty years earlier) and argued that 'the album is not obscene in law however hateful you may find the lyrics'. Describing rap as street journalism, and relying on a defence of rap as 'realism' rather than on the wider aesthetic implications of hip hop music culture,[32] Robertson stressed that 'it is told in a crude language, almost a patois. It is often very bitterly sarcastic and rude and will appear to our ears rude and crude, but there it is, all part of the experience. It tells it like it is.'

After hearing the album the magistrates agreed with Robertson that the record was not obscene because it was not likely to deprave and corrupt, titillate or excite sexually and they had not believed beyond reasonable doubt that the record would corrupt. Following the judgment, the group arranged to distribute two thousand posters across the United Kingdom in a defiant gesture towards the censorship campaigners. The posters included the citation of Article 19 of the Universal Declaration of Human Rights, 1948, which states that 'everyone has the right to freedom of opinion and expression', and the terse injunction to fans and supporters to 'Express Yourself'. The NWA spokesperson told the music press in Britain that the group 'feels vindicated at the judgment passed and the poster campaign is an attempt to hammer the point home that freedom of speech is absolutely essential in a democratic society'.

The eventual 1992 USA presidential election campaign victor, Bill Clinton, publicly criticised 'raptivist' Sister Souljah, formerly of

Public Enemy, for her explanations of why Los Angeles youths decided to kill whites instead of each other, comparing it to the white racism of neo-Nazi David Duke. But as Ted Swedenburg (1992) points out, the US media did at least give rappers attention in the wake of the LA riots because no one else seemed to have a sense of what was going on in South Central Los Angeles. Civil liberties supporters were, and are, divided about many of the rap censorship cases. Some have argued for constraints on rap records which are viciously racist, while others fear that the anti-obscenity campaigners will achieve censorship through intimidation, and that, particularly, the labelling controversy in America will be imported wholesale to the UK. David Toop, musician, writer and historian of rap (1984, 1991 and 1992) has covered the NWA case in all its sordid details and wondered aloud and in public whether the individual perpetrators of tales of brutal male lust and other illiberal anecdotes (like Eazy-E of NWA) are really worthy of the protection of anti-censorship libertarians. On the other hand, Paul Oldfield and Simon Reynolds have argued (1989) that the new sub-genres of pop such as rap and hip hop (as well as Belgian New Beat, and other body pop styles) – not to mention old 'cock rock' styles such as heavy metal – have embarrassed the liberal press because they are masculinist, macho, sexist and sometimes fascistic. These writers claim that ignoring the genre's allure for participants, or else welcoming new, softer or feminine versions, has often been the tactic adopted in the liberal (music) press. For Oldfield and Reynolds, however, such music can in fact be a deconstruction of machismo. As they say, 'for many of us the music lets us get under the skin of maleness and see its impoverishment, numbness, struggles with weakness, and dangers, as well as the exhilaration'. The case of Madonna,[33] and especially her controversial picture book, *Sex*, has been cited by some gay and lesbian writers[34] as a more progressive use of provocation and invitation to the state to censor art through law than, say, rap music. In this instance the sadomasochistic nature of many of the photographs in the book – though clearly part of a marketing strategy by Madonna and, once again, the company Time-Warner, to sell both book and accompanying album *Erotica* – was frequently applauded for its erotics of popular culture, for its attempt to open up cultural space for progressive libertarian ideas and practices.

Censorship of art forms is also carried out by regulation of the institutions and technologies of the media as well as in cultural forms such as music or photography. For example, radio, as both an institution and a technology, has had more than its fair share of control, surveillance and discipline. In Britain, popular music radio has increasingly become part of the establishment since Radio 1 began in 1967 as an alternative to the popular pirate radio of the 1960s.[35] As a national institution to rival the Royal Family such radio has achieved the power of longevity even if this has meant that its DJs have grown older with it. As John Peel, by some way its most radical broadcaster for almost thirty years and initially plucked like many of the first Radio 1 employees from a 1960s pirate station, has openly said: 'I'm an overweight, balding, middle-aged man with a rather unsuitable pony tail and four children. It'd be absurd if I claimed to be going out raving every night, out of my head on Ecstasy. People'd think "Poor old tosser!" '

However, for most British Conservative Party politicians, even in general those who are young enough to be part of what has been termed 'the rock and roll generation', Radio 1 functions as a negative reference point when planning broadcasting legislation. BBC Radio Lancashire's cult three-hour Sunday afternoon programme, *On the Wire*, was axed – and then reprieved – in late 1991 by a newly appointed manager at the station in favour of what BBC Radio's local research said listeners wanted; that is, news, sport and information. *On the Wire*, with, according to the correspondent in *The Independent* newspaper who campaigned for its reinstatement, its 'bizarre mix of jazz, blues, reggae, hip-hop, thrash metal and pretty well anything that catches the ear of its protean presenter, Steve Barker, and his coterie of corespondents' did not fit this agenda. However, widespread protest, particularly through the newsprint media, forced a change of heart and a subsequent reinstatement, though the show was eventually moved to a Thursday evening slot.

Debate at a national level has continued amongst the regulators of British radio about what is not pop music as defined – after much debate in the House of Lords during the passage of the Bill – in the Broadcasting Act, 1990, which included attempts to control broadcasting time for different kinds of music. The difficulty revolved around defining what the (then) Thatcher government meant by a radio service offering music 'wholly or mainly . . . which is not pop music'. Their Lordships defined pop music – the object to be avoided – as 'rock music and other kinds of popular modern music which are characterised by a strong rhythmic element and a reliance on electronic amplification of their performance'. The phrase 'other kinds of popular modern music' posed problems for the Radio Authority's receipt of bids for the licences to run national commercial radio frequencies. For Simon Frith (1990) this equation of 'pop' and 'rock' in the Broadcasting Bill was a setback for the music industry, making a national FM rock station along American lines unlikely. Mark Fisher, then Labour Shadow Minister for the Arts and Media, who perversely described 'pop music [as] rock music without the sex or soul', (1988) remarked about the Home Secretary and his plans for the Broadcasting Bill, that despite rhetoric of deregulation, privatisation, individualism and freedom:

> we are living in an increasingly centralist and authoritarian state. We have a government which will use every parliamentary, political and legal means to strangle public debate, as *Real Lives*, the Zircon tapes, and the *Spycatcher* saga chillingly remind us. Coverage of current affairs is shrivelling; meanwhile, titles like *The Sun*, *The Star* and *Sunday Sport* carry so little hard news that they scarcely deserve to be called newspapers. It is indeed a bleak media landscape.

'Tabloidisation' will be noted in section 6 as a pernicious and pervasive process in the 1980s and 1990s in many countries in the world, and, as Robert Chapman (1994), has demonstrated, 1990s pirate radio has experienced a 'law and order' crackdown from the state as unlicensed stations are taken to court and equipment confiscated leading to the closure of stations.[36] In Britain, officers of the Department of Trade and Industry (DTI), the department of government responsible for policing the airwaves, has stepped up the clampdown, armed with legal changes introduced at the beginning of 1991. Whilst the new Broadcasting Act – an Act supposedly designed to increase choice for radio consumers – was being debated in parliament the DTI brought in amendments to existing legislation. Changes in the Marine etc. Broadcasting Offences Act, 1967, the legislation introduced by the Labour government in the 1960s to control the pirate radio of the day, were introduced to strengthen the state's powers against unlicensed broadcasters such that, at least in the eyes of some commentators,[37] the rules to regulate illegal radio stations exceed the government's regulatory powers over even, for example, drug smuggling. The changes to the 1967 Act certainly mean that pirate radio stations broadcasting offshore can now be raided outside the twelve-mile limit of United Kingdom territorial waters.

In this section we have extensively studied the terrain of what I called at the beginning the 'law of art' or the 'state and the art'. In the remainder of Part II of the book we need to put in place the foundations for what an aesthetics, and an erotics, of law might comprise in the 'law and popular culture' field as it is now being (re)drawn in the various practices of popular culture and legal regulation which have been covered in this section.

Notes

1 Especially as host of a popular British soccer fan phone-in programme on BBC Radio 5, called '606' after the show's starting time. On Mellor's championing on the programme of civil liberties for football fans, see Adam Brown (1993).
2 See Davies and Hung 1993; for a brief assessment of this book, and this particular international legal field, see Redhead 1994a.
3 See Hirst and Kingdom 1979/80.
4 See for a major collection of essays, Frith 1993, and my review of the book, Redhead 1994b.
5 See Rosemary Coombe's fascinating work on this, especially Coombe 1992.
6 See the Ph.D. thesis of Ronald Moy at the Institute for Popular Music, University of Liverpool, 1993, for the development of the concept of 'enginician', a hybrid of musician and engineer.
7 See the extensive work on copyright, authorship, sampling and contemporary dance music done by Hillegonda Rietveld, a Ph.D. student at the Manchester Institute for Popular Culture. Rietveld, as a recording artist and DJ herself, has lectured on this research to students on 'law and popular culture' courses using samples as illustrations.
8 See chapter 11 of their book.
9 See Amanda Chandler's postgraduate work on law, popular culture and deviance in this area at the Manchester Institute for Popular Culture. She has coined the term 'intellectual joyriders' to describe this general (mainly young male) fraternity of computer outlaws.
10 See for example Nicholson Baker, *Vox* (Granta, London, 1992).
11 See for example Leslie Dick, 'Mintel 3615', in Marsha Rowe, (ed.), *Sex in the City* (Serpent's Tail, London, 1989).
12 See especially chapter 4.

13 See Beaumont 1993.
14 See Robertson 1980, on the socio-legal implications of the Williams Committee report.
15 See Pratt and Sparks 1987.
16 See Brown, 1980 and Kuhn 1984.
17 See Petley 1984.
18 See Taylor 1987.
19 See Ellis 1980.
20 See Willemen, 1980.
21 See Stanley 1993.
22 See Cheney 1993.
23 See Brown 1981, and Eckersley 1987.
24 See, for instance, Milne 1987.
25 See Avkigos 1993, on Cindy Sherman's art, for instance.
26 See Mort 1980, and Cousins 1987, for excellent and incisive commentaries on Foucault's work on this, together with many sections of James Miller's brilliant biography of Foucault (1993).
27 See interviews with Eazy-E, as well as Ice-T and Ice Cube and many more West Coast American rappers, in Brian Cross's extensive contextualisation of Los Angeles rap and hip hop (1993) where they discuss their views on 'gangbangin' ', homophobic attitudes and issues raised by the politics of the black underclass in the United States and the random gang violence and hatred for the police. Another Los Angeles rapper, Snoop Doggy Dog, whose LP *Doggystyle* was a huge international commercial success, was charged with murder in 1993 in what police maintained was a gang vendetta (see Reed 1993) after he and his bodyguard chased an old adversary and killed him. To counter these taken-for-granted connections, see black gay film-maker Isaac Julien's programme for *Arena* on BBC television, transmitted on 12 February 1994, which looked critically at the masculinist and homophobic attitudes of much rap, and also reggae, music.
28 See Leigh 1993.
29 See Robertson 1979.
30 See Robertson and Nichol 1992.
31 See Robertson 1989.
32 See Cloonan 1993.
33 On the phenomena of Madonna fandom, especially amongst women, see Turner 1993.
34 See Frank and Smith 1993.
35 This process is bound up with the retrospective 'selling of the 1960s' as Robert Chapman (1992) has convincingly shown.
36 See Goldberg 1991.
37 Goldberg 1991.

5

The love of law and popular culture

When I was first drafting this book an academic colleague – and series book editor – suggested that the life history of the word 'postmodernism' might itself be at an end, judging by the number of manuscript proposals (including, at that time, this one!) landing on his desk which highlighted the term. His point was that the concept had become so ubiquitous, at least in the 'rarified' world of academic literary theory and cultural studies book writing, that it no longer made sense – even if it ever had – to use it. For some writers[1] references to 'post-postmodernity suggest it is a time already past'. Others have of course noted, with some cynicism, the multiplicity of referents of the label 'postmodern' over the last decade or two;[2] indeed they have argued that such multiplicity is one of the defining tenets of a 'postmodernist culture'. My academic colleague was, however, signalling something potentially more significant. His caution was meant to draw attention to the fact that, in the 1990s, we may be – in a variety of senses – past the 'post'. This section, and indeed the whole of this second part of the book, explores the possibilities and problems of moving towards a post-postmodern 'law and popular culture' – in terms of its form both as an academic discipline, or intersection of disciplines, and as a regulatory regime. In this section I want to look at the bare bones of what I call an aesthetics of law, or the art of law, and also an erotics of law.

First let us consider the important and stimulating, but ultimately flawed, project of establishing a 'postmodern jurisprudence' (or for that matter a 'postmodern criminology'). This has been forged in much the same way that a 'postmodern sociology' has been attempted in the wider social and human sciences where debates have raged over whether what is being created is, or could be, really a sociology of postmodernism rather than a 'postmodern sociology'.[3] Indeed the precarious status of sociology, both as an academic discipline and as an institutional area of study which is reflected in faculty and departmental organisation within higher education, has been under severe threat during the period of these debates. As I argued in the context of the sociology of pop and rock music and the sociology of youth in *The End-of-the-Century Party*, the attack on sociology in the 1980s came from a pincer movement of New Right thinking (Margaret Thatcher's famous pronouncement that there is no such thing as society) and one apparently nihilistic brand of postmodernism which proclaimed that we were now living the 'end of the social'. As is also extensively argued in *The End-of-the-Century Party*, the urgent task may be not simply to go with the flow of these two arguments 'against sociology' but to begin to reconstruct 'the social'.[4] The writer most controversially associated with the theoretical development of the term 'postmodern', and the thesis of the 'end of the social' is Jean Baudrillard. However, there are many reasons to see this connection as mistaken, as Baudrillard's most critically sympathetic commentator in sociology, Mike Gane, has skilfully shown (1991a and 1991b). Indeed, Baudrillard has even argued in interview with Mike Gane (1993) that:

> Everything that has been said about postmodernism was said even before the term existed . . . So what to do? To want to disassociate oneself from it, to say that I am not a postmodernist, is still to say too much because it is a contradictory opinion and therefore defensive, and I don't want to go along that road either. . . . So I have nothing to say about this because I say, and I know this from experience, even if I prove that I am not a postmodernist, it won't change anything. People will put a label on you. Once they have done that it sticks. There are ways of getting rid of a problem. People just label it and in my opinion this is not really a process of . . . how shall I say . . . it's not really a fraud . . . let's not exaggerate, but there is something that is not very clear in all this. There are perhaps areas where the term 'postmodernism' may mean something to the extent that people claim to belong to this, perhaps in architecture. But as soon as it is clear that the term adds nothing new it is best to let go of it. But it'll be around for a while. It has been around for a long time already.

In Part I of this book I have explored the connections between legal discourse and what can in many ways be described as its 'other', the discourses – and practices – associated with popular culture, and to a lesser extent postmodernism. In a textbook specifically entitled *Postmodern Jurisprudence* which attempts to explore 'the ethics and politics of reading the law in an age of uncertainty' (Douzinas, Warrington and McVeigh 1991),[5] the authors say that in their view:

> Postmodernism must neither replace one set of certainties with another, nor create a new series of bipolarities. Indeed there is the danger that its levelling function can lead to an affirmation of the banal culture of the 'global village' and a celebration of simulacra that 'learn from Las Vegas' and seek to imitate it. The political imperative of postmodern deconstructive readings is to remain critical and oppositionist, and to challenge any orthodoxy that a complacent and affirmative postmodernism may wish to reimpose. If law is politics by other means, a deconstructive reading of law means other politics.

Quite apart from the reappearance of the very bipolarities they seek to avoid – the 'bad' postmodernism, 'complacent and affirmative" must presumably be defined in opposition to the 'good', 'critical and oppositionist' postmodernism – the 'banal culture of the global village' is not a cultural condition which can be set aside so easily. As Baudrillard has shown in his more aphoristic, less sociological, books, *America* (1988), *Cool Memories* (1990) and *The Transparency of Evil* (1993), events like the New York marathon and the Heysel stadium disaster and activities such as driving and flying across the roads and deserts of the United States can be pregnant with both banality and significance. Despite many insights and provocative analyses (such as the analysis of modern copyright law which we noted in section 4) the style, language and choice of examples for analysis of 'postmodern jurisprudence', at least as far as it is represented in this particular textbook, display an uncritical contempt for popular culture, or the 'banal culture of the global village',[6] and a reinstatement, and moreover a re-emphasis, of the traditional high and low culture divide in jurisprudence.

For a so-called 'postmodern jurisprudence' to come into being the object 'postmodern law' has had to be described. One theorist who has sought, amongst others such as Douzinas, Warrington and McVeigh, to apply 'postmodern perspectives to law and society' Anthony Carty,[7] is a scholar better known for his work in international legal studies. Carty has edited a diverse collection of essays on just such a theme, concentrating on what might be seen to be the strictly post-structuralist – rather than the postmodernist – theme of the death of 'man' and the Enlightenment age of reason (Carty 1990). For some writers who are critical of this tendency to merge two types of theory – jurisprudence on the one hand and what Douzinas, Warrington and McVeigh (1991) have called 'postmodern deconstruction' on the other – the fact that 'post-modern jurisprudence delights in parody, derision, ridicule and drama'[8] is a problem. This appears to be because some of its adherents characterise postmodernism as 'melancholic, the symptom of our contemporary malaise' rather than a source of humour or jokiness. However, there seems no reason why the 'lie back and enjoy it', celebratory, 'end-of-the-century' mood/mode of some of the writing about the postmodern condition should not co-exist with a sense of melancholy, nostalgia and a sense of loss which other writers, especially those like Peter Goodrich (1986) who are concerned with the transformations of 'law and modernity' and 'postmodern jurisprudence', have seen in the condition of postmodernity.

Jean Baudrillard argued (1991) controversially[9] in the context of the 1991 Gulf War that what is in dispute is the 'corpse of war' and that, 'where everything transforms itself into a 'virtual' form, we are confronted with a virtual apocalypse'. Indeed, after safe sex and safe war, there is no reason why there cannot be thoroughgoing expansion of such theoretical concepts as safe law, or for that matter, as we have already noted, 'virtual law'. In the times of 'the *last* law', 'safe law', would come after 'dead law'. It is a sign of the struggle over the corpse of law. As with an obsession with the end of or death of the 'social' or politics – or history, or ideology – there has been considerable energy expended in jurisprudence over the last decade on considering and sometimes asking questions about whether this is an era of the 'end of law'[10] or the 'death of law'.[11] An 'end-of-the-century jurisprudence' – as I would rather term the project of theorising post-postmodern 'law and popular culture' – has to take into account, though not necessarily follow, the formative conditions of what I have labelled here 'post-structuralist jurisprudence' where we find a different reading of the law, or 'Reading (of) The Law', as Peter Goodrich puts it (1986). Here there is an alternative use of legal methods and techniques which are completely alien to those of legal positivism but derive directly from a literary theory and linguistic philosophy which itself encountered and resisted the dominance of postivism in linguistics since the nineteenth century.[12] One lesson from such explorations has been, controversially, that the study of law is in essence a literary pursuit – a claim which has its problematic repercussions in the fields of postmodernism and popular culture where, as again Peter Goodrich (1990) notes, the image is increasingly dominant, and where indeed 'the contemporary cultural terrain is defined almost exclusively in visual terms, including the display, the icon, the representations of the real through the camera's eye, captured on videotape and given in the moving picture' as Chris Stanley has persuasively argued (1993b, see 1993d). This vision of postmodern law, or 'panic law', is of law in a 'post-literate' culture, already in the here-and-now.

In the search for the mapping of a 'postmodern jurisprudence', cartography has proved to be one theoretical resource. In drawing one such draft version of a map of the law, Boaventura de Sousa Santos (1987) suggests that 'the relations law entertains with social reality are much similar to those between maps and spatial reality' and that, if law is one way of imagining the real, 'many unresolved problems in the sociological study of the law . . . may be solved by comparing law with other ways of imagining the real'. For other critical legal scholars, such as Allan Hutchinson (1988), it has been necessary to draw on various art forms (films, plays and so on)[13] in order to jolt the 'reader's traditional expectations about the style and presentation of academic argument' and to provide 'critical challenges to received jurisprudential wisdom'. As Alan Hunt points out in a wide-ranging review (1988) of the project, Hutchinson 'experiments with a range of stylistic genres' where 'the reader encounters a one-act play, snatches of sheet music, poetry, cartoons and other devices' such as mock judgements, film reviews and fake folk tales. But Hunt essays the view that this 'new style' of postmodernist legal scholars is 'an immediate give away and provides an apparent excuse for unreconstructed legal academics to turn the page in search of "some real law" '. Such styles, though, in fact are not particularly new and have frequently varied more widely to include more pervasive aspects of contemporary popular culture such as rock and pop music[14] as well as alternative aesthetics of 'high' culture.[15] A consideration of the place of aesthetics in the law itself has also been undertaken on occasions[16] but the full exploration of the possibilities of an aesthetics of law has not developed, since law, as usual, would seem to be the last discipline for such a tendency.[17]

Rather it is the 'deconstructive turn' which writers like Hutchinson have taken which has come in for most criticism, either from cautious but empathetic critics like Hunt (1990), from hostile writers as Paul Carrington (1984), or even from those like Douzinas, Warrington and McVeigh (1989), who write from the heart of what they call 'postmodern deconstruction' itself and have employed just such techniques of emphasising the 'performative' mode in their own work. For Douzinas, Warrington and McVeigh, in particular, much of what Hutchinson does with his tools of deconstruction is praiseworthy but these critics worry that by committing himself to tell 'law's stories' in a new way 'he may make himself vulnerable to the criticism that the problems of left-wing politics are not solved simply by a sudden conversion to story telling'. They also worry that by urging a deconstructionist project as against an Enlightenment one he, and others, can fall prey to the charge that this constitutes the setting up of yet another unjustifiable binary division which, as they point out elsewhere (Douzinas and Warrington, 1987), any 'postmodern decon-struction' worthy of the name would seek to dismantle. Hutchinson's deconstruction is in some ways both over-faithful to the work of Jacques Derrida and also, eventually, more like an instance of the strategy for 'trashing' aspects of legal precept and practice which Christopher Norris (1988) has attributed to the vast bulk of the critical legal studies movement in general. Critical legal studies, in Norris's view, can be further accused of nihilism, irra-tionalism and 'an interpretative free-for-all with not the least regard for the standards of argumentative rigour and truth'. The problem with the criticism of Norris lies in the difficulty of the concept of 'truth' in law.[18] Deconstruction, in legal studies and literary theory, proposes a reading of texts, especially philosophical (or jurisprudential) texts, drawing on literary criticism. Its essential thrust is to debunk the claim of a distinction between 'philosophy' and 'literature'. That claim is that, in a legal context, jurisprudence, in its widest sense including statutes and judicial decisions, gives a scientific insight into 'truth' which legal literature, or a play or film about the law (say, Oliver Stone's *JFK* or Jim Sheridan's *In the Name of the Father*), does not. The latter relies on the 'play' of language which the philosophical text does not in the claim that the two are distinct. Deconstruction rejects the terms on which these binary divides – law and literature, fact and fiction – are constructed. As with other binary divides, deconstruction rejects their basis, arguing that philosophical or scientific texts are just as subject to the play and 'figurality' of language as 'literary' ones.

As already noted, I want to argue, contradicting other common claims, that it is a 'post-structuralist' rather than a 'postmodern jurisprudence', which has been constructed so far in legal and cultural studies, and, moreover, that elements of a 'postmodern jurisprudence' briefly discussed above (virtual law, dead law, safe law, panic law) are extremely difficult to represent within conventional discourse. Attempts like that of Stephen Pfohl (1992) to write a postmodern sociology of law and criminology utilise all kinds of different creative writing styles, (still) visual images, and so on, without fully conquering the summit of Jean-François Lyotard's mountainous question for art theory and art history about representing/theorising what he called the 'postmodern condition' – in other words how to 'present the unrepresentable' (Lyotard 1983). Feminist work in this field[19] has either concentrated on showing how post-structuralist thinking can conquer problems encountered in earlier formulations of the sociology of law and deviance, or else constructively and innovatively appropriated the term 'postmodernism' as the late Mary Joe Frug has done (1992) in forging new directions in legal feminism. Writings on the history and politics of post-structuralism itself[20] have tended to ignore these new forays into legal theory, which are often inspired by semiotic analyses[21] and expounded on the pages of literary theory and cultural studies journals such as *Textual Practice* and *Cultural Studies*. Semiotics, or semiology, the science of signs first developed in structural linguistics in the early twentieth century as a result of the work of Ferdinand de Saussure, has contributed significantly for many years to the theorisation within legal and cultural studies. Especially noteworthy is its demonstration of the bond between the 'signifier' (the thing, word or object) and 'signified' (meaning). Deconstruction, as part of a general movement in the human and social sciences known as post-structuralism, has emphasised the instability of the relation between signifier and signified, which contrary to earlier histories of semiology and structural linguistics is less fixed than previously thought. The philosophical basis of deconstruction, namely that any text (legal, literary, cinematic, architectural or whatever) inevitably undermines its own claims to have a determinate meaning, is obviously radically subversive. The criticism of determinate meaning, and consequently the 'presence' of truth – and the heralding, and celebration, of multiple interpretations which deconstruction brings – has fundamental consequences for formalistic legal logic and language, as it has in non-legal fields.

Where post-structuralist writings on law have moved on to a terrain which could be important and fruitful is when trying to theorise 'desire' and particularly the complex and fraught field of legal desire, or 'the desire of the law' as Mark Taylor (1990) has called it. As Neil Duxbury (1989) has pointed out, the 'idea that pyschoanalysis might be of use in the study of law and legal activity is by no means anything new', but new directions have been forged in recent years by excavating (and translating) work of continental jurisprudence and critical philosophy as well as psychoanalysis itself. Sigmund Freud's seminal nineteenth-century theorising has been developed and extended in a number of legal contexts. The work of the French classicist, psychoanalyst and legal theorist Pierre Legendre[22] has received attention and translation[23] in the pages of journals such as *Law and Critique* and the *International Journal for the Semiotics of Law*. As with much other work on psychoanalysis in the context of critical theory since the early 1970s, it is Jacques Lacan's (1977) radical re-reading of Freud which Legendre plunders, especially the idea of 'the law' represented by the father figure. Lacan, who influenced the structuralist Marxism of Louis Althusser in the 1960s and 1970s, and consequently many CCCS versions of cultural studies, theorised the formation of the infant as subject in language through unconscious processes.

The writings of Gilles Deleuze, friend and biographer of Michel Foucault, and Félix Guattari (Deleuze and Guattari, 1983 and 1986) have also inspired much of this renewed investigation of psychoanalysis and law in the 1980s and 1990s in both casual references – for example, the subtitle 'nomadic masks' to Peter Goodrich's study of the 'languages of law' (Goodrich 1990) – and more detailed exposition of their translation of Lacan's idea of 'the law' into the context of the modern state which has an 'unconscious' subject to 'the law'.[24] The difficulties which were initially perceived to exist with application of Lacanian psychoanalysis to realms such as social and legal theory in the 1970s are still evident; particularly those which relate to the problems of analysing non-human subjectivities – such as state, law and corporate forms – with the aid of Lacan's theory of the three orders underlying the relations of human beings, namely the Symbolic, the Imaginary and the Real. Moreover, the concentration on psychoanalysis is by no means of itself a radical break with general legal theories from earlier in the century. American legal realism has long had psychological and psychoanalytical foundations (albeit of a behavioural kind) and is particularly prominent in the 'back to pyschoanalysis' ouevre. A conference in New York in September 1993 on legal studies and psychoanalytic and deconstructive modes of theorising self-consciously focused itself around the theme 'Law and the postmodern mind', explicitly bringing out the American Realist reference point in the persona of Jerome Frank whose book *Law and the Modern Mind* was first published in America in 1930.[25]

This pyschoanalytic interest in law brings us back to theorising the realm of the legal subject, the 'subject-in-law', which in the context of Marxist theory of law we explicated in section 4. However, this theoretical analysis clearly needs to move beyond mere notions of subects, and subject positions. Without a theory of legal desire, theorising of subjects-in-law remains in a pre-structuralist or structuralist quagmire. A notion of the 'sliding signifier' always already underneath the signified is helpful here. The idea of the body-in-the-law which we mentioned in conjunction with cyber law in section 4, could be re-posed in this context, as well as, to mistranslate Pierre Legendre, 'the love of the law'. This conceptualising of a 'love of the law' – or what I would want to call an erotics of law – is plagued with dangers not least because of the continuing feminist and gay critiques of both Freud and Lacan. The 'love of law' here is not necessarily a heterosexual one; a 'queer law'[26] is just as possible but it needs to be predicated, as Jean Baudrillard (1990) sees it, on the idea of 'seduction'. It is the law 'in drag' – translaw alongside transaethetics and transsexual – which is the most suggestive interpretation one can make in this context: it is a kind of cross-dressing of law. The question of sexual difference and the law is relevant in terms of both legal education and legal and cultural theorising. In terms of their construction in the academy – and especially the law school – both masculinity[27] and femininity[28] have been subject to rigorous theoretical scrutiny.

Legal strategy for feminist change has further been the subject of renewed efforts[29] to conceive difference and the law in a way which moves beyond the difficulties of a 'jurisprudence of equality' (claims for more equal pay and sex discrimination legislation and its more efficient enforcement, for instance) and on towards a 'jurisprudence of difference'. Much of the most pertinent criticism in this field has come from literary theory, and especially the fertile ground of French feminism and post-structuralism; the work of Hélène Cixous (Cixous and Clément 1987) and Julia Kristeva[30] is significant here, as is that of other long time contributors to the debates such as Luce Irigaray.[31] Justice, from these theorists' position, can be conceived as a kind of 'feminine writing' of the body-in-law. Writers like Elisabeth Bronfen (1992) have pushed further on with the questions of the body, femininity and aesthetics into the realm of 'death'. Jean Baudrillard, too, in his most well received book in France in the 1970s (1993), has conceived of symbolic exchange and death as the terms to interrogate the interlocking struggle betwen the symbolic and semiotic orders. It is the realm of 'death' of the law which we now need to consider in order to probe the individualised, marketised world where youth, deviance, law and popular culture meet on the mean streets of the major world cities of the late twentieth century.

This section has briefly ranged over various post-structuralist, postmodernist and other cultural studies approaches to law – a postmodern law review – and by implication moved beyond the formal boundaries of disciplines such as jurisprudence and criminology. In the 'death' and deconstruction of the subject areas themselves we find ourselves at the end(s) of law: post-law.

Notes

1 For instance, Smart 1993.
2 See, for the longest list of meanings of postmodernism I have yet discovered, Hebdige 1986/7.
3 Mike Featherstone, Professor of Sociology at the University of Teesside, and his colleagues on the journal *Theory, Culture and Society* and the acclaimed 'Theory, Culture and Society' book series, for the publishers Sage, are especially responsible for the high quality of the theoretical debates around this difficult issue.
4 It may of course be the notion, and widespread feeling – in Britain at

least, after many years of New Right government – of an 'end of politics' which is the key to this question. Interestingly, Martin Jacques, editor of the now defunct journal *Marxism Today*, argued this point in the rhetoric of a title of a lecture at the University of Salford in February 1993 implying that it is time to 'forget society' but to reinstate 'politics'. This task is the subject of future work but it is worth noting Paul Hirst's (1994) rejection of the necessarily 'postmodern' trajectory formed by the collapse of Stalinist Soviet-style socialism in the East in the late 1980s and the simultaneous exposé of the flaws of free market capitalism in the West.

5 See also Douzinas and Warrington 1991.

6 The work of the Australian cultural and feminist critic Meaghan Morris has made far more critical sense of Baudrillard, banality and cultural studies than anything written about these issues in 'postmodern jurisprudence'; see Morris 1988.

7 See Carty and Mair 1990; see also Milovanovic 1992.

8 As one such critic of 'postmodern jurisprudence', Neil Duxbury (1991), has put it.

9 See Christopher Norris's (1991) tirade against Baudrillard in the context of the Gulf War. Although Baudrillard asks for much of what he gets, Norris's defence of the absolute political correctness of certain philosophical positions – and the incorrectness of others – seems to me to be untenable.

10 See O'Hagan 1984.

11 See Fiss 1986.

12 See Goodrich 1988a.

13 Indeed Hutchinson's book title, *Dwelling on the Threshold*, is an adaptation of a Van Morrison song, 'Dweller on the threshold' from the *Beautiful Vision* LP. It can be added as another instance of plunder from, and allusion to, popular music culture which legal studies has engaged in; see section 2 of this book for many more examples.

14 See the use of Chuck Berry's 'Roll over Beethoven' as a title for a seminal CCLS article written by Duncan Kennedy and Peter Gabel (1984).

15 For instance, Peter Goodrich (1990) has included, as the final chapter of a formally conventional jurisprudence book, a detailed analysis of the work of artist and film-maker David Walliker, including black and white illustrations. The chapter originally appeared as a catalogue, with stylish black cover and colour illustrations, for the artist's exhibition.

16 See Douzinas, Warrington and McVeigh 1991; the chapter in the book which specifically deals with legal theory and aesthetics was also included in an issue of *Theory, Culture and Society*, one of the very few articles published in the journal to cover 'postmodern jurisprudence'.

17 The 'Theory, Culture and Society' book series for Sage has featured some important interventions which have pushed sociological debates 'past the post' into territory which fully recognises the importance of aesthetics: see, especially, Scott Lash and John Urry's development of the concept of 'aesthetic reflexivity' in Lash and Urry 1993. Further, Simon Frith has for a number of years prepared the ground for his work on an 'aesthetics of popular music' rather than a 'sociology of rock'.

18 That is, if we transpose deconstruction from art and literary criticism to law, the 'truth in law' is the focus of what we would come up with. Jacques Derrida, in particular, has written about – for instance – the 'truth in painting'.

19 See Young 1988.

20 See Easthope 1988.

21 See Goodrich 1988.

22 See Duxbury 1989, Duxbury 1990, and Goodrich 1990.

23 See Goodrich and Warrington 1990.

24 See Stanley 1993a, and Duxbury 1989.

25 For a detailed and rigorous excavation of American legal realism, see Neil Duxbury's forthcoming account of American jurisprudence (1994).

26 See especially the theoretical work on this of Les Moran (1993) and Stephen Whittle (1994).

27 See Collier 1991.

28 Bottomley *et al.* 1987.

29 See Bottomley and Conagh 1993.

30 See Lechte 1993 for a good summary of Kristeva.

31 For the best introduction to all French feminist literary theory including Cixous, Irigaray and Kristeva, see Moi 1985.

6

The ends of law and popular culture

In this penultimate section of the book I want to consider in more depth what some of the concepts referred to in section 5 might look like when applied in the realm of 'post-law'. What does the diction and fiction of legal desire, and the desire of the law, actually mean in a post-AIDS world where, now that there is more danger than ever in social as well as sexual intercourse, the 'seduction' of law is almost an end in itself? In the ends of law, the entrails of law, we search for an answer to questions blurred by the phenomenal speed of change in popular and youth culture.

In what is increasingly being seen by cultural critics as 'postmodern culture' – 'post-literate' culture to many other writers – in which, as we have seen, a visually-oriented mass media loops soundbites, public opinion survey results and news-as-entertainment for a global audience glued to television and video screens, it may seem somewhat perverse to concentrate in this section on law, crime and deviance as concepts which are now best portrayed in the form of fiction. The novel form, though, is still a tempting style to use for the interrogation of 'rough justice' cases and real-life crime incidents. John Williams (1994), for example, adopts some elements of this style to investigate the horrific murder of a prostitute in Cardiff in 1988. Williams (1991) also argues in a fine combination of travelogue and multiple book review of American crime writers from James Lee Burke through James Ellroy to Sara Paretsky that the genre of crime fiction is most telling about the 'hard' underbelly of the American dream where in the 1990s 'the have nots are still making something new' (the hip hop culture of the streets) while the 'haves are looking back fondly at the 1950s'. In Britain for both youth and crime fiction this is equally true. In 1964 Charles Hamblett and Jane Devenson wrote *Generation X*, a 'classic' pulp novel on youth. In 1976–7 Billy Idol and his mates took the same name for another blank generation, the mid-1970s punks, and named their band Generation X. The same title is taken for Douglas Coupland's first novel (1992). *Generation X* is subtitled most aptly *Tales for an Accelerated Culture* and was first published by St Martin's Press in the United States appropriately in 1991, the year of the Gulf War. The Gulf War was the first major world conflict to be televised live in America, and constituted a savage video-war game event with horrific consequences in terms of the forgotten dead, injured and ill victims. As we have seen in section 5 it was marked by what Jean Baudrillard controversially and playfully called a 'reality gulf' (1991). *Cosmopolitan* magazine described *Generation X* as 'truly a modern-day *Catcher in the Rye*',[1] an epithet which is becoming the dubious literary equivalent of the popular music critics' never-ending search for the 'new punk'. 'Punk' novels like Jess Mowry's *Way Past Cool* (1992) and *Rats in the Tree* (1993) and Richard Price's *Clockers* (1992) are chock full of (black) American street language which is now the most obvious global sign of the times, a signification of both juvenile delinquency and youth culture whether it be in Barnsley, Barbados or Baltimore. Youth itself is still the most prominent sign of crime and of 'law and order' needing to be mobilised to combat a rising social problem.

The novel form – rather than, as in the most recent past, ethnographies or sociologies of youth crime and deviance – has in fact become the main means of representing modern juvenile deliquency and a supposedly disappearing myriad of deviant youth cultures in the 1990s. Contemporary youth, or lifestyle, novels and portraits by Michael Bracewell (*The Crypto-Amnesia Club* 1988), *The Divine Concepts of Physical Beauty*, 1991 and *The Conclave* (1992)),[2] Martin Millar (*Milk, Sulphate and Alby Starvation* 1987, *Lux the Poet* 1988, *Ruby and the Stone Age Diet*, 1989 and *The Good Fairies of New York*, 1991) and Gordon Legge (*The Shoe*, 1989 and *In Between Talking about the Football*, 1991), amongst others, have told us more about the 'feel' of contemporary youth culture than hundreds of pages of sociology of youth and criminology and sociology of deviance studies from the formally academic arena, though there are just as many bad youth culture novels[3] as there are dull, out-of-date and misinformed sociologies of youth deviance and delinquency. A similar trend, where novels or novellas have tended to displace social science accounts of law, crime and deviant lifestyles, has occurred in the United States with what Elizabeth Young and Graham Caveney, in the best book to be written about this field (1992), call 'blank generation' fiction (after punk musician Richard Hell's song). This 'downtown' writing is exemplified for some critics by Jay McInerney (*Bright Lights, Big City*, 1985, *Story of My Life*, 1989 and *Brightness Falls*, 1992) and Bret Easton Ellis (*Less Than Zero*, 1986, *The Rules of Attraction*, 1987 and *American Psycho*, 1991), though many would also perceive these to be essentially works of 'yuppy' (young urban professional) fiction; both the social status and the main characters of the books of McInerney and Ellis certainly meet the requirements of this photofit. On the other hand it could be argued that neither punk nor 'yuppy' culture is the historical basis for fiction of this kind; it is Anthony Burgess's *A Clockwork Orange*, first published in 1962, and made into a much-banned/withdrawn film in the early 1970s, which has perhaps most strongly influenced such writing for three decades now.

A more accurate description of the work which Young and Caveney celebrate so enthusiastically would be 'post-punk'. Writers such as Kathy Acker (*Blood and Guts in High School, plus Two*, 1984, *Don Quixote*, 1986 and *Young Lust*, 1989) have attracted epithets of this sort since the late 1970s. *City Limits* praised her book *Don Quixote*, from the mid-1980s, as 'a chart of American gender decay and schizoid post-punk culture' long after the term 'post-punk' had become fashionable to describe the music of bands such as The Gang of Four and The Mekons in the late 1970s and early 1980s. This issue of 'gender decay' is precisely, however, the focus which holds together the disparate writing in this area and gives it its undeniable cutting edge. Much of the content of these texts, especially Acker's, reads like pornography, though the declared intention is to disrupt the notion of the 'I' of the author in as many ways as possible, thus unsettling the comfortable reader's 'position' from which the text is read. *American Psycho*, by Bret Easton Ellis (1991) which brought the author major opposition from publishing houses, feminist groups and assorted representatives of the 'new puritanism' is a classic example of this genre. Despite the fact that, as Chris Stanley (1993b) shows, the reading of such a text can be recuperated for radical opponents of sexism, there is considerable danger in the direct copying of a pornographic style of writing to hold up the genre to attack. Where this is more obviously used to effect in 'post-punk writing is in gay men's fiction where authors like Gary Indiana (*Rent Boy*, 1994) and Dennis Cooper (*Frisk*, 1992, *Wrong*, 1994 and *Closer*, 1994) portray a hedonistic resistance to the terrors of an AIDS-dominated agenda and take on the mantle of Hubert Selby and William Burroughs in romping through the brutality and tenderness of contemporary culture and crime, law and deviance. Indiana even begins his book *Rent Boy* with a definition of 'pornography'. These books are certainly not for the faint-hearted reader with their concentration, especially in Dennis Cooper's *Frisk*, on murder, mutilation, drugs and predatory sex.

There have been numerous instances of fiction *about* blank generations in Western culture since the 1950s, though Ryu Murakami's 1977 Japanese youth novel recently republished in English (*Almost Transparent Blue*, 1992) is a reminder that the United States and Britain by no means had a monopoly. The phenomenon of youth culture novels is a familiar feature of pop culture, as *Generation X* (both 1960s and 1990s versions) reminds us. The postwar decades, at the very least, have been littered with 'pulp' and 'serious' (itself virtually a 'low' and 'high' brow distinction) within the popular culture genre of youth culture fiction. From Richard Allen in the 1970s, whose cult youth trash novels are now being reprinted in collected volumes (1992, 1993), via Nik Cohn (*Ball the Wall*, 1989) in the 1960s and 1970s to Colin MacInnes (*Absolute Beginners*, republished 1980) in the 1950s there have been endless re-interpretations of a 'pop' writing style depending on whether realism, modernism or postmodernism was in vogue at the particular juncture.

What marks out the current crop of youth culture novelists like Douglas Coupland is firstly that they are writing about the generation younger than they are – the 'twentysomethings' in *Generation X* and, in his follow-up novel, *Shampoo Planet*, (1993) 'global teens', or those who grew up in the Reagan and Thatcher years and who 'came of age' in the early 1990s. The second distinction is that writers like Coupland are being published in an era where the decline of sociology of youth and sociology of deviance, as a discipline, and as a genre of writing, is fast becoming a cliché. As we saw in section 5, sociology as a whole and these sub-disciplines in particular have been under attack from neo-liberal market economics on the one hand and varieties of 'postmodern' theory on the other throughout most of the late 1970s and 1980s. Social science's claim to an 'authenticity' in representing, for instance, varieties of youth cultural street style and argot as was the case in earlier decades is, for some commentators, fast becoming a cruel joke. One reason for this is simply age; participant observation work on youth culture needs to be done by someone of either gender who is roughly twenty-three to thirty-three years old. Though it is by no means the only methodology appropriate, it is necessary if not sufficient to be of this age bracket. As postwar sociologists have got older, and research grants to hire postgraduate students have become fewer, the inevitable has happened and social scientists have tended to lose touch – as both fans and professionals – with what 'your actual young person' does and thinks today.

Such problems of the process of ageing do, however, not always constitute the whole story: *Youth and the Condition of Britain* by John Davies (1990) could appear in print many years after its research period had finished (around 1980) and consequently have virtually nothing to say about youth in even the Reagan and Thatcher years, let alone those of Bush, Clinton and Major. In as far as youth studies have been about the narrative of 'my generation', it is essentially a generation which grew up in the wake of World War II that has formed its subject; moreover, this is a 'rock and roll' generation for whom pop (music) culture in particular was more significant than Sega and Nintendo computer games, club culture[4] or joyriding. Although there is a continual mass media proclamation of 'the death of pop', or a posing of the omnipresent question 'Is pop dead?', as we saw in section 2 forty- and fiftysomething academic attempts to utilise such rock culture for pedagogic purposes can seem so outdated in a fast-moving pop culture context as to appear farcical, however well intentioned. Nevertheless, the question of 'authenticity' and 'inauthenticity', and how much we can believe the teller of law, crime and deviance tales, is at the heart of the difficulties of contemporary judgements on (re)presentations of law and transgression in popular culture.

Let us consider some examples of popular youth and deviance fiction. For instance, *Tribes* (1992) is Alexander Stuart's follow-up to the much-acclaimed *The War Zone*, a debut novel which wove strands of narrative into a seductive portrayal of incest, adolescent anger and social (dis)order and was described by Anthony Burgess as a 'pungent, shocking book'. His other published work has been in the area of children's fiction (his own young son died of cancer), film criticism and screenplays. Stuart was executive producer of Nicolas Roeg's *Insignificance*, a role which clearly prepared him for the hero-as-film-maker (or film-maker-as-hero), Nick Burns, in *Tribes*. One review compared it both to the low-culture street violence action of Richard Allen's skinhead book series and the high-culture hi-tech, ultra-violence of Burgess's *A Clockwork Orange*, with science fiction writer J. G. Ballard's *Crash* and Gordon Williams's *Straw Dogs* sandwiched somewhere between. Stuart, through Nick Burns, tells us that 'Violence is the sex of the nineties. It has replaced sex. We need the thrill of violence to turn us on.' This 'made-for-reviewer' soundbite may seem much too facile, as is much of the cardboard cut-out plot and its one-dimensional cast of characters, but there is an important sense given here of the terrain of transgression.

The portrait of English football hooligan violence that we are given in *Tribes*, especially, is (literally) unbelievable. It is as unrepresentative of contemporary youth culture as another supposedly non-fictional – albeit beautifully written – account given by *Granta* arts magazine editor Bill Buford (Buford, 1991) in *Among the Thugs*, his much-discussed story of English football hooligan behaviour in the 1980s and 1990s. Both Buford and Stuart employ, and re-promote, an 'image' of the English football hooligan which is so stereotypical and so manifestly displaced in historical time (late 1960s to early 1970s not 1990s) that the reader of both books is forced to conclude that the hidden subject matter is the 'pride and prejudice' of the intellectual with regard to the masses and popular culture, as John Carey has put it (1992).

Undoubtedly, both Buford's and Stuart's concern is essentially the connection between the historical question of the 'mob' and 'the crowd' and the respectable citizenry, and also the contemporary sexual politics of a fast-fragmenting male sexuality and the plurality of masculine identities on offer in the (post)modern world. As the dust jacket by publishers Chatto & Windus puts it, Stuart's novel 'explores the connections between the aggression of the streets and the intimacy of the bedroom, between violence and tenderness, between the urge to dominate and our need to be loved'. The book itself barely lives up to such billing. The main female character (presumably representing 'tenderness') is thinly drawn and the major focus of the narrative is really maleness (for which read 'violence') and the middle-class film-maker's obsession with working-class thuggery in the shape of 'the Neck', the neighbourhood football hooligan. As in Buford's writing, though, the interest in Stuart's novel lies in the literary skills employed rather than any anthropological knowledge and insight into research methods which might have been gained. Where the book does work is not in the fake authenticity of the street dialogue but in the pointers it makes to the tapestry of the postmodern media milieu which Nick Burns – and, perhaps, the real author and film-maker, Stuart, himself – inhabit. This world is one where constant re-invention of character, story-line and personality (the self, the subject) is permanently possible; all human action, including violence and tenderness, is translatable into 'image' – in this case the film, also called *Tribes*, which Burns is making and which produces the real-life violence of a Wembley soccer stadium 'shoot' as the Neck and his boys realise the lower orders' potential for savagery in the big crowd scene. Nick Burns emerges as a good, old-fashioned comic strip hero who comes back fighting and even wins the girl (and her child) in the end. White male angst at the end of its tether pervades a book supposedly set in an end-of-the-century London whose geographical context could have been anywhere in the United Kingdom – as long as the date was some twenty years ago.

On the other hand, two books where setting is, in a literary sense, all-important are Jess Mowry's *Way Past Cool* and Richard Price's *Clockers*. In another Chatto & Windus 1990s 'youth' novel, Mowry relates the tale of a gang of black teenagers fighting to survive the ravages of the free market in the post-Reagan United States. Published in the year of the Los Angeles riots, this narrative of delinquency and drift is appropriately set in the black ghettos of Oakland, California. Mowry lived in Oakland, we are told, and has been 'a drug dealer's bodyguard, made his living from collecting scrap metal, and worked in a centre for street kids'. His place in the thick of things thus duly established, Mowry goes on to produce a script which sounds like a Public Enemy twelve-inch single. It captures male, young, hip hop, black culture superbly. A gritty realism with a no-holds-barred dialogue is extremely effective in this (a)morality play of guns 'n' drugs 'n' rap. This is the tale of tough roughs in the best tradition of youth delinquency novels stretching back to Pinkie in Graham Greene's *Brighton Rock* (young gangster as hero seeks sympathetic reader having found empathetic author).[5] In California in the 1990s, modern-day Pinkies pit themselves, and their individualist ideologies, against unleashed and unfettered (market) forces.

Mowry is as adept at anthropology as novel-writing and the skatekids come alive in their tight black 'tees' (faded to grey), ragged Levi 501 jeans with holes and big trainers; this is a carefully observed world where 'cheapo' K-mart watches jostle for position with more expensive Nike and Reebok footwear. LA gear is everywhere today as black street styles[6] metamorphose all over the place in the global shopping malls where both white kids and the fashion houses raid the 'street' for inspiration. But in Mowry's novel this sits alongside a macho street 'wisdom' which divides life into a hard part and a girl part: 'girls come at you sideways most times', writes Mowry. School culture – this youth culture is *under* eighteen – is presented as a looped series of rites of passage, an experience, of law and laws, without end. The main reason for this is that modern life, as Thomas Hobbes from the vantage point of the seventeenth and eighteenth centuries predicted, can only be nasty, brutish and short, a veritable war of all against all, when de-regulation of guns is so widespread. The street dialogue spits with gun fire – Uzi, .24, .387, .45 – as well as school essays, switchblades, television and cars and girls.

If we want to know about growing up young, black, male and underclass in the American (East or West) Coast city at the end of the twentieth century, this would, at first sight, seem to be the novel to buy. In any case the film rights have already been sold and it will be coming to 'a cinema near you' following hard on the heels of *Colors, New Jack City, Boyz 'N' The Hood, Straight Out of Brooklyn* and *Do the Right Thing* (with their soundtracks culled from some of the rap artists mentioned in section 4) and many others of the future. But therein lies the rub. The subject matter of *Way Past Cool* is already familiar to us through a whole decade of products from the cultural industries like popular music, video and film. This is a kind of post-realist realism, not just a post-rap novel. However 'authentic' his account of street life turns out to be, we already know what to expect; we've been down these mean streets before. Rap may be, commercially, a 1980s and 1990s phenomenon but mostly it's the same old story (man as 'baaad' and woman as 'feminine', though the 'black woman as strength of the community' mythology gets a plug, too, as it does in some of the best female rap groups like Salt 'n' Pepa) in the same, if rearranged, setting in Mowry's hip (hop) novel.

The location moves to the East Coast – a 'city just outside Manhattan' – of the United States for Richard Price's *Clockers*. The title is street slang for a crew of corner-boy cocaine dealers and much of Jess Mowry's terrain is covered here. Price is author of the best-selling book-as-film-script *The Wanderers*, his first novel from the early 1970s reissued in paperback in 1993 after the commercial and critical success of *Clockers*. He has since become one of Hollywood's most successful screenwriters (he wrote *The Color of Money* for instance), and frankly it shows. You can picture Robert de Niro reading between so many of the lines here; sure enough Universal pictures are already filming *Clockers* and de Niro stars in two new movies based on Price's screenplays. The problem is that this tells us more about American-based multi-media conglomerates' post-1991 Gulf War preoccupation with US male street violence and machismo than about the nuances of what is really going down on the 'other side' of the world's bright lights and big cities in the 1990s. Price's chosen genre-mix is youth novel meets American detective fiction with, like Mowry, a keen eye for detail of street style and language, and for the increasing social problem of widespread (crack) cocaine use and organised crime activity in its production, distribution and exchange. The reader, though, may get the impression that, unlike Mowry, Price does not really have the passion for this fashion – at least not any more. It reads too much as if, for the author, writing about street youth culture has become a well-paid job of work (albeit one to carry out efficiently and effectively) and, as long as the punters lap it up, the feeling seems to be, well it's only (pulp) fiction, not social documentary, anyway. For commercially successful black social documentary fiction at the cutting edge Victor Headley's (1992, 1993) 'realism' cleans up comprehensively on British streets where Jamaican-based 'Yardie' gangs, guns and crack cocaine stalk their prey – increasingly, it seems, the (tabloidised) mass media and (black) independent publishers.

Fiction, though, is not the only representation of the end(s) of law. Journalists investigating the streets, the bedrooms and boardrooms of the post-law world's major cityscapes have occasionally also dug deep into the connections between youth and deviance. *The Other Side* by Ruben Martinez (1992) is an example of another publisher with impeccable left-wing credentials, tackling the 1990s youth, rock, popular culture, law, crime and politics field. There is no fiction here (though there is poetry) but plenty of rattling good narratives of investigative journalism from the frontline – and in view of the earthquake in 1994, fault-lines – of Los Angeles (where Martinez is a staffer at *LA Weekly*), Mexico City, San Salvador, Tijuana and Havana. In the book there are some rich accounts, not to mention fine photographs, of Latino graffiti art, hip hop culture and Mexican 'rocanrol'. Fragments of Chicano-influenced youth styles pervade this book, which takes its style from the journalism of war reporting; these are stories and reflections from the 'battlefields' (in some cases, literally). The book also has a striking colour cover montage of a picture of a graffiti mural by Los Angeles artist Cooz and a photo of Mexico City rock band Maldita Vecindad y Los Hijos del Quinto Patio. The picture of the 'other' culture – Hispanic-speaking America – which Martinez builds is certainly powerful and helps to underline a more complex racial view of, say, the LA riots in 1992; Los Angeles is, for example, the centre for rock music from Latin America and Spain as well as a capital of hip hop. Racism is a constant up-front theme of the book. There are snapshots of 'rock against Stalinism' in Cuba, contributing to a re-think of the culture of the 1940s and 1950s as well as the post-1959 revolution in that country, but the Los Angeles vignettes are the best part of the book. *The Other Side* is an excellent complement, in that sense, to Mike Davis's *City of Quartz* (1990). Both books predate the riots which followed the Rodney King jury's (first) not guilty verdict (King was the black motorist beaten up by a group of LA police) but anybody who read either of them and could not see the conflagration coming is plain blinkered. There are AIDS stories, hip hop street culture graphics (with a Jungle Brothers rap tape as soundtrack) and a sharply razored feel of the casual violence and random shootings of Los Angeles ghetto life in Martinez's book. The detail in, say, the graffiti art case studies is fascinating, too: the reader learns intricate details of bombers, writers and tags and how LA streetstyle rivalled New York wildstyle, as well as the East versus West competition in Los Angeles itself. But this not all there is in/on the 'other side'. The traditional musics of Mexico and Cuba are subjected to some

astute analysis (Tijuana brass is resurrected, theoretically, in the
face of US-promoted Barbie culture) as is the place of 'rock' music
in imperialist and colonial culture. The piece about the book's
cover stars, the rock band from Mexico City, is a fine example of
Martinez's reporting: we learn of the postmodern mix of fashion
styles – U2, James Dean and film-maker Luis Buñuel's barrio kids –
and the musical genres of danzon, mambo, rap ska, funk and rock
that Maldita Vecindad embrace, and find out how youth subcultures in Mexico contribute to the breakdown of barriers.
Martinez argues that Mexico City intellectuals (half-jokingly) say
that postmodernity actually originated there five hundred years ago
with the Conquest and its clash of radically different sensibilities. It
is noteworthy, too, how the Catholic church, Marxist left and the
Mexican government all agreed with the proposition that Mexican
youth were corrupted by the Protestant, decadent and indivi-
dualistic North America.

For a short course in postmodernism and the youth culture
novel, Douglas Coupland's *Generation X*, already mentioned, is
outstanding. Unlike McInerney and Easton Ellis's work, it is an
anti- or, more accurately, post-yuppy novel for the 1990s.
Coupland has produced, in form and content, a text which is a
sheer delight. It skates the surfaces of the depthless, flattened,
accelerated (like the cover of a book by arch-theorist of the politics
of speed, Paul Virilio)[7] – televisual culture which gives it its subtitle
with a good deal of deadpan wit and a parodic, self-conscious
writing style. A superb pop art cover wraps a fast-moving tale of
three twentysomethings (two men, one woman) whose defining
characteristic, or sign, is that they come to represent a time after the
decade of unadulterated material greed, deregulation and
computerisation of financial markets and the B-movie talking head
as leader of the world. A glossary of useful contemporary terms
appears in the wide margins of the text (sometimes with illustrations and slogans) to explain everything from 'Brazilification' (the
widening gulf between the rich and the poor and the accompanying
disappearance of the middle classes) to 'the Tens' (the first decade
of new century). This is a (love) story of a (hyper)real phenomenon,
a youth culture which has, eventually, become solely an advertising
fiction. This is, in truth, the potential fate of the subjects (and
objects) of youth culture as global cultural industry as we leave the
twentieth for the twenty-first century. As with Jean Baudrillard's

concept of disappearing 'social', there is no escape here; re-invention of youth culture as such is impossible while market(ing) forces lump together almost anyone between five and fifty-five as susceptible to youth cultural commodities (including youth novels) and – simultaneously – contribute to the decline of the pheno-menon itself, making it nothing other (neither less nor more) than the collective imagination of the industry. Appropriately, Coupland's work of fiction finishes with the findings of a (factual) telephone poll of six hundred and two people aged eighteen to twenty-nine in America in June 1990 as part of an appendix called 'Numbers'. Youth has disappeared, finally and irrevocably, as it recognises itself only in the returns of public opinion polls. The penultimate chapter of *Generation X*, though, is the real key to the (dead) future of youth culture; like Jean Baudrillard (1988) the author proposes to miss out the last decade of the twentieth century altogether as the 'event' has already taken place, or feeling perhaps that it will never in fact take place. The title of this penultimate chapter is 'Jan 01 2000'. Welcome to the World of Hyperlegality. Welcome to Accelerated, Post-Literate Culture. Welcome to the New Millennium . . .

Notes

1 Alexander Stuart's first novel *The War Zone* (1993) drew similar reviews.
2 See also Bracewell's excellent introduction to intimate photographic study of Morrissey as a pop and youth icon by Linder Sterling – known simply as Linder (1992). In addition, see Bracewell *et al.* 1988 for a collection of one of Bracewell's earlier novellas, 'Missing Margate', alongside work by Don Watson and Mark Edwards.
3 See, for instance, Robert Elms' (1989) flawed attempt to write a youth culture novel, a surprising failure in view of his many incisive pieces of 'street journalism' on precisely the terrain of youth and youth culture in magazines such as *The Face* in the 1980s.
4 It is contemporary dance culture rather than rock music which, since the mid-1980s, has defined youth and pop culture. Turning Dick Hebdige's notion on its head, we can today trace a move from 'subculture to clubculture' in both academic writing on youth culture and street life itself over the last two decades.
5 Like Burgess, Greene is an often 'unacknowledged source for 'punk', 'post-punk', or 'downtown' writing on deviance, law and youth. That

musical and performance styles sometimes also derive from such novels is testified by John Lydon (alias Jonny Rotten), singer in the 1970s punk band The Sex Pistols, who claims *Brighton Rock* as a considerable formative influence (see Sawyer 1994).

6 The major exception is 'grunge' which as a musical and a fashion style has a long history beginning with the delayed transatlantic trip of English punk in the 1970s to America in the 1980s; see, for the best history so far, Arnold 1993.

7 See, for instance, Virilio 1991.

7

Unpopular cultural studies

Part autobiography, part commentary on its own construction, part reference book for students on a variety of humanities and social sciences courses, part series of book reviews, this book has explored the fertile deconstructed terrain where legal theory, deviance and cultural studies collide: a zone where there are no longer any recognisable academic disciplinary boundaries or border/lines. It has told, for the first time, the story of the birth and subsequent history of 'law and popular culture'. This term now designates both a field of study within the rapidly marketising academy and a regulatory regime in the 'real' world. The areas covered in the book have been wide and diverse: sport, the 'arts' (high and low), popular music (rock, dance, pop and so on), heritage, pop tourism, myriad youth cultures, information technology (hackers, phreakers, tappers) and all of the mass media. The histories and studies excavated and exposed here suggest the existence of an identifiable condition in the 'fin de millennium' legal and cultural theory which is only partially analysed in the debates about postmodernism, postmodernity and postmodernisation. The new disciplinary terrain of popular cultural studies can be shown to diverge from previous attempts to conceptualise law, deviance and culture but also to be firmly rooted in the rich – if flawed – legacy of contemporary cultural studies and critical legal studies. In this final section of the book I want to convey some of the excitement, and anxiety, of being involved in pioneering in a small way some aspects of popular cultural studies and to tie together the history of the popular/unpopular couplet which parts of the preceding sections have reconstructed and unravelled and followed to the maelstrom of our own accelerated popular culture. 'Popular' and 'unpopular' are fast replacing 'legal and illegal', 'normal' and 'pathological', 'straight' and 'deviant' as the key terms of the debates for what once were the unreconstructed domains of academic disciplines such as jurisprudence, sociology of law, sociology of deviance and criminology.

Simon Frith, Director of the John Logie Baird Centre for Film, Television and Popular Music at Strathclyde, Stirling and Glasgow Universities, has argued (1992) that the attempt to rescue pop consumption from the 'condescension of high cultural snobs, on the one hand, and from the despair of folk purists on the other' has floundered. For Frith a popular and accessible journal that took popular culture seriously would be an excellent idea but in his view those that have emerged, like *The Modern Review*, have failed miserably.[1] They are in Jim McGuigan's (1992) phrase part of a widespread 'cultural populism'. As Frith, writing against cultural populism, says 'cultural populism means anti-intellectualism (obvious in the pages of *The Modern Review*), endless carping displays of common sense . . . and a peculiar faith in the authenticity of "ordinary" experience'. In my view, taking popular culture seriously means developing arguments about what is 'popular' as well as what is 'cultural'; that means, as Frith notes, that different 'criteria are needed: criteria in which unpopularity is given its proper measure of respect'. Frith, in a much-discussed essay (Frith and Savage 1993) with popular music critic Jon Savage[2] has also pointed to the ways in which 'studying popular culture has become a method of uncritical celebration', turning into a populism which defends 'low culture' and dismisses high culture, the reverse of what has occurred for a much longer period of time in circumstances where high culture has dismissed the low. *The Modern Review* itself has responded to its critics with a predictable mixture of confused populism and haughty disdain, especially in the form of articles and talks by its editor, Toby Young (1993) culminating in an Oxford Union debate[3] which revived the kind of elitist, high culture attitudes and prejudices which in fact underpin much of the ethos (though not all the contributions) surrounding the magazine. Many of *The Modern Review*'s responses to the backlash against cultural populism consisted of wondering aloud why the publication has not been welcomed within the academy in general and cultural studies[4] in particular. The more interesting question is why so much media space has been given to the vanity publishing of Julie Burchill, a former *New Musical Express* writer in the mid to late 1970s and for many years afterwards a right-wing columnist

for *The Face* and, eventually, for several British national news-
papers as well as a pulp novelist.

As Jon Savage (1993) has claimed, 'that old chestnut, the High
Culture, Low Culture division' is integrally related to a 'wider
picture: a concentrated, right wing attack against the freedoms of
popular culture, of which pop music is the spearhead'. He suggests
that the increasing claims that 'pop is dead' both have no bases, in
that pop music is as good as ever – he cites on his own behalf
techno, Suede, Nirvana and the Pet Shop Boys in 1993 – and are
part of a shorthand for a pervasive, puritan and moralistic
nostalgia for a pre-1960s golden age; what British Prime Minister
John Major described in 1993 as 'back to basics'. For Savage:

> Implicit in the attack against pop music is the recognition of its
> power. Pop is the least censored of all the mass media industries:
> within it, performers and audiences find a space to express how they
> really feel, who they really are . Now that the idea of the popular has
> become elided with the populist (the uncritical acceptance of market
> values, a reductionist view of humanity), pop music reminds us that
> the extraordinary, the valuable, the forward looking can be popular
> as well. This is important. We live in a time when our politics and
> culture harks back to a non-existent past, just when new ideas are
> most needed. Within pop's sounds and modes of organisation –
> pluralistic, decentralised, fluid and flexible, inclusive – you can hear
> what Britain is like and how it could be. We denigrate pop music
> constantly, and we do so at our peril.

In Savage's view these are the reasons why pop is not 'nowhere' in
the scheme of values which privileges 'the novel' as 'the highest
form of artistic activity'. Angela McRobbie has further argued
(1991) that a moral panic about the threat posed by pop culture to
critical seriousness is also a moral panic about art and popular
culture, in particular the problem of low culture being infused with
the aspirations and intentions normally associated with art. The
debate over whether 'Keats is more important than Bob Dylan', or
vice versa, is clearly about formations of national identity, too. The
images of Britishness and Englishness, to take one example of
popular culture and nationalism, can be gleaned from popular
music culture just as much as, if not more than, from Shakespeare,
Blake, Burns, Turner and Purcell. Simon Biddell and Mark Fisher,[5]
for instance, authors of a book on popular music and the postures
of high culture, choose five figures (all white and male!) from pop

music history to try to reveal a national consciousness that is distinctly contrary to that of high culture: Syd Barrett (early Pink Floyd), Ray Davies (The Kinks), Morrissey (The Smiths), Mark E. Smith (The Fall) and Neil Tennant (Pet Shop Boys). In the USA, cultural studies has become the 'most controversial and contested terrain of the 1990s' according to Michael Berubé, an academic writing in the *Village Voice Literary Supplement* (1992) about a large conference organised at the University of Illinois at a time when even the representatives of the American liberal intelligentsia such as Michael Medved are arguing that popular culture no longer enriches. Such liberal-turned-new-puritan critics instead write books (Medved, 1993) claiming that the entertainment industry assaults the American (and other) people's most cherished values, and corrupts their children.

It seems that what was earlier in the twentieth century a modernist avant garde is now to some extent a (post)modernist popular cultural fragment defining itself against a 'populist cultural studies'. 'Popular culture' in any case is not a coherent academic subject, whatever the British Open University Popular Culture (U203) course in the early 1980s might have proclaimed. It is in fact, as this book has sought to show, a much-contested, and fast-changing, terrain. Popular – as opposed to contemporary – cultural studies is, however, a discipline which constitutes as its (fleeting, always disappearing) object the field of 'popular culture'. How far an affective involvement is felt, or how uncritical or critical (Hall 1981) the stance regarding the object in question is, is of course, crucial. In the work of John Fiske (1989), for instance, there frequently appears to be a celebration of virtually any object of desire in popular culture (from Madonna to blue jeans). For Stuart Hall (1981), whose own work was so important to the CCCS and the British Open University Popular Culture course, 'popular culture' is a term which designates a domain of struggle and is of interest only as such, not for any other reason.

The Manchester Institute for Popular Culture at the Manchester Metropolitan University was set up precisely in order to produce a basis – theoretical and material – for a popular cultural studies to emerge. It was created out of a federation of the Unit for Law and Popular Culture in the School (at that time Department) of Law and the Centre for Urban and Cultural Analysis in the Department of Social Science when the University was still a polytechnic. Its focus was to promote theoretical and empirical research in the area of contemporary popular culture, but with a long historical sense of the 'contemporary'. This was to be done both within the University and in conjunction with local, national and international agencies. The Institute runs a number of postgraduate research programmes with a particular emphasis on ethnographic work, strongly favouring participant observation for much of the field work. A collective research effort and regular presentation of work in pro-gess to research seminars and publication in the form of working papers and book series are viewed as essential.

Documentation of 'popular memory' is also seen as vital through oral history and archival work. The Manchester Institute for Popu-lar Culture archive is – mainly – a unique collection of print magazines. It was Simon Frith who in the introduction to Craig McGregor's collection of essays *Pop Goes the Culture* (1984) coined the term 'print addict', and when I first read the essay I realised that the label fitted the project of archiving fan magazine culture perfectly. For as long as I could remember I had obsessively collected the trivia – though it was always deadly serious stuff to me – of fan magazine culture. These were the first of literally thousands of fan magazines I was to expend a small fortune on over the next three decades. Many years later, at the end of the 1980s when I was setting up the Unit for Law and Popular Culture, which eventually spawned the Manchester Institute for Popular Culture, I had amassed a large amount of this ephemera. A book published in 1991 by Wordsmith, independent publishers in Manchester, called *Football With Attitude* drew heavily on this collection, and the Institute held a very successful exhibition in November 1991 based on Richard Davis's unique fan 'vox pics' taken especially for *Football With Attitude* and *The Passion and the Fashion*, a sub-sequent book on football fandom in the New Europe. It also drew on the ever-increasing memorabilia archive. Alongside an equally cumbersome and obsessively filed record and tape collection in my home, the archive consisted of collected print media, in the sense of books and journals, but always, also, the 'debris' – the fan magazines which researchers could never find in any library but which constituted the raw material for what, at least in my own mind, could be already labelled as 'popular cultural studies'. This, as this book has demonstrated, can be defined to include the study of fandom and fan obsession. The archive, thanks to the Institute's hard-working researchers and many donations from kind-hearted fanzine editors and equally generous collectors, has expanded massively since the first clear-out from under my bed. The idea was to give the donated material a good academic location and, in the process, to assure better, more interesting, research into popular culture from all over the world.

In addition to the print archive, the Manchester Institute for Popular Culture has a video and audio tape library and a newscuttings collection. Bona fide researchers are welcome to come in and use the archive. The Institute was set up in order to recreate some of the conditions of the 1960s and 1970s Centre for Contemporary Cultural Studies at the University of Birmingham[6] but in the much changed climate of the 1990s a large number of differences had to be recognised. The theoretical developments in the social and human sciences since the emergence of CCCS are perhaps of most critical importance and call into question much of the neo-Marxist foundations of the 'Birmingham School'. Collective research work and ethnographic methodologies, however, are a lasting, and highly significant, legacy of the CCCS, and it is now vital to retain them if a lasting and modern cultural studies is to develop.

What might a popular cultural studies look like then in comparison with contemporary cultural studies? High and low culture can be studied together,[7] though the key term 'popular' should be used instead of 'contemporary'. It is the 'regulation' of popular culture which marks out the terrain of study, where popular and unpopular clash. But 'popular' is resonant as much of 'pop' (music, culture) as some kind of 'low' or 'mass' or anti-high-culture formation and care has to be taken with its use. Popular cultural studies parodies and pastiches previous contemporary cultural studies just as pop parodies and pastiches its own socio-musical history, as I argued at length in *The End-of-the-Century Party*. Such 'play' with the past of cultural studies is not denigrative but reverential and playful at the same time. The Manchester Institute for Popular Culture, for example, produces a working papers series called *Working Papers in Popular Cultural Studies* designed deliberately to draw attention to the excellent stencilled working papers (and their subsequent collection into edited books) produced by CCCS, and to pay a certain homage to an important and rich heritage in cultural studies, but also to highlight the battleground over the popular/unpopular couplet which shifts the direction of the interdisciplinary field I have described and analysed in this book. The 'Manchester' in the title of the Institute for Popular Culture was used to register both a distance from Birmingham and its cultural studies 'school' of the 1970s and the shift in pop cultural sites (of production and consumption) which determined that Manchester (as 'Madchester', 'Badchester', 'Gunchester' and 'Gaychester' in the tabloid headlines) became a global focus for pop and youth culture at the end of the 1980s and beginning of the 1990s. The use of *for* Popular Culture – though it is widely misprinted or pronounced as 'of' – in the Institute's title signified a commitment to some, at least, of the progressive connotations of 'popular culture'. The book series, entitled 'Popular Cultural Studies', which the Manchester Institute for Popular Culture negotiated in 1992 with Ashgate Publishing (publishing house for Gower, Arena, Avebury, Edward Elgar and Scolar), was conceived as similar to the Hutchinson series which for many years collected together the work of the CCCS and its associates in book form, often highlighting revised versions of the original stencilled papers. *Rave Off* and *The Passion and the Fashion* were both deliberately organised around such principles, sampling the best of the Unit for Law and Popular Culture working papers series which predated the Institute working paper series.

However, there were other influences which were being acknowledged in this new deviance pop/law aesthetic. For instance, the subtitle to *Rave Off* was *Politics and Deviance in Contemporary Youth Culture* self-consciously excavating the popular Penguin paperback series of National Deviancy Conference books (one of which was called *Politics and Deviance*) consisting of the accessible, up-to-date and innovative edited conference papers which I discussed in section 3. Other 1960s and 1970s publications which were signified in the new series in the 1990s were the long running 'Penguin Special' series and the Routledge direct editions. The books in the 'Popular Cultural Studies' are prepared camera-ready at the Institute which makes it relatively easy to produce and publish the series so that the titles can capture the speed of change at the times when the studies were written.

The first texts I myself wrote in this field in the mid-1980s were again designed to utilise strategies of irony, pastiche, parody and humour in a version of a low modernist[8] aesthetic. *Sing When*

You're Winning, for example, was produced at Pluto Press where excellent books on rap and women in pop[9] had been published. Designed by Tony Benn, who had worked on avant-garde journals such as *ZG*, the book self-consciously fits into the style of these two previous popular culture books in a direct act of simultaneous plunder and praise, theft and celebration. The writing style of the book was intended as an over-the-top pastiche of the rhetorical writing style of Julie Burchill, whose own collection of vitriolic prose had just been published as *Love It or Shove It*. There was a deliberate cultural and political purpose, too, in the aesthetic strategy used to portray the 'findings' of sociological and deviancy research which had been done over a number of years at Manchester Polytechnic, and the University of Warwick. Eugene McLaughlin and I had written an essay[10] in *New Society* at the beginning of the 1985/6 British soccer season, in part to counter the celebration of the actions of soccer casuals by Toby Young, (later to become part of Julie Burchill's coterie), in *The Observer* in an article entitled 'Saturday Afternoon Fever'. Another reason was to intervene in academic and policy debates about crime, law and football hooliganism which were then dominated by those responsible for the research produced at the University of Leicester, later to become the Sir Norman Chester Centre for Football Research.

Access to the mass media was gained through these strategies, a platform which would not have been available then if we had pursued a policy of producing conventional academic reports. This strategy was redoubled for the production of *Football With Attitude*, which came packaged in a brilliant open-out pop art cover which was essentially a pastiche of Peter Blake's celebrated LP cover for the Beatles' *Sgt Pepper's Lonely Hearts Club Band* in 1967 but with an appropriately 1990s overload of images which tumbled off the edges of the paperback, front and back. The research reported on in the book, whose text was partially a parody and theft of *Sing When You're Winning* itself, covered the football legislation to control fandom, crime and deviance in Britain in the 1980s and early 1990s. The study of legal and social regulation contained in the two books showed how regulatory regimes around discourses of 'football hooliganism' produced, and continue to reproduce, cultures of 'defence' rather than 'resistance', as the CCCS work in this field had contended over a decade earlier.

But the commodification and aestheticisation of this argument about 'juridification' and soccer culture is in itself crucial; the presentation of the arguments was inseparable from the content – the medium really was the message. Dada, surrealism, pop art, post-punk 'trash' aesthetics, photo-journalism, modernism and postmodernism, were all drawn upon to construct a certain 'product' which would be consumed in a host of different ways (some of which were, inevitably, at odds with the 'intentions' of the author as recounted here – hooligans themselves were known to consume at least some of the supposedly ironic passages of the book with some glee). What was common, too, to these two books as commodities was not just the beginnings of a (pop) aesthetic of law but an erotics of law. The focus of much of the two books was the 'masculinities' produced and regulated in the disciplining of soccer and youth. To emphasise this the launch for *Football With Attitude* was held at Manto – a bar in the heart of Manchester's gay village[11] – where the photographs from the book adorned the walls. Soccer culture is still one of the most male-dominated, and 'machismo'-oriented popular cultural formations. The irony of such a venue was consciously intended; as a young man growing up in the 1950s and 1960s in Britain I recall how being interested in (and good at) soccer meant a defence from much of the vicious male bullying at school, and in the street, which was invariably targeted at those who failed to live up to the stereotypical behaviour of 'being a man'.

In the preceding pages what I have set out, then, is a way forward in theorising, and instituting, a popular cultural studies – which might just as well be termed 'unpopular cultural studies' in view of its emanation from the bowels of jurisprudence, sociology of law and sociology of deviance as well as cultural studies. What readers of this present book might conclude is that there has been something of a complacent time lag in the development of legal studies of popular culture. The argument in this book has been that developments in popular cultural studies – and to an extent in 'popular culture' – implode legal theory (which itself has absorbed media culture). What needs to be worked towards is law not simply as an art form but as a popular cultural form. This requires new tools of analysis to understand it. I hope that the pages of this book provide the outline of a route towards such a destination.

Notes

1 See Patrick Wright's (1992) coverage of the debate about the journal and its cultural politics, quoting Simon Frith and others' views.

2 Jon Savage was the first Visiting Fellow at the Manchester Institute For Popular Culture, the Manchester Metropolitan University, from September to December 1992. The essay, written with Simon Frith, was originally published by the Institute as *Working Papers in Popular Cultural Studies* no. 2, before being eventually published as an article in *New Left Review* in 1993 (Frith and Savage 1993), and is forthcoming in a collection of some of the series of Manchester Institute for Popular Culture working papers edited by Steve Redhead, Derek Wynne and Justin O'Connor entitled *Popular Cultural Studies* and to be published by Blackwell, Oxford. The essay was originally commissioned as a review essay by the *London Review of Books* but turned down on what the Institute regarded as utterly spurious grounds. After protesting to the magazine about the decision, the Manchester Institute for Popular Culture published it in 'samizdat' form. Simon Frith delivered the first annual lecture to the Manchester Institute for Popular Culture in May 1993, when in a talk entitled 'Instituting popular culture' he warned against the dangers of cultural populism.

3 Reports of this were again carried in the British national press by Patrick Wright (1993).

4 See Antony Easthope's thoughts on this lack of reception for the magazine within academia (1993a and 1993b). Easthope's argument was formed partially as a response to a vitriolic panel discussion on popular culture at the Cornerhouse arts centre, Manchester, in December 1992, when he was a member of the audience and Toby Young, Paul Morley, Jon Savage and Rosalind Brunt (formerly of CCCS, now of the Centre for Popular Culture at Sheffield Hallam University) were members of the panel.

5 See Simon Biddell and Mark Fisher's book *Lay of the Land*, forthcoming. Antony Easthope discussed with postgraduates at the Institute some of the issues around national identity and popular culture at the Manchester Institute for Popular Culture open postgraduate seminar in January 1994, in his talk 'Thinking about (English) national culture', the text of which forms part of a forthcoming book project.

6 See John Clarke – who has, as was made clear in section 3, for many years been a major player in the 'ballpark' of Contemporary Cultural Studies – on his personal, autobiographical view of the working and theoretico-political conditions of CCCS in the early 1970s in *New Times and Old Enemies* (1991), especially chapter 1. The book

features Clarke's reflections on the New Right dominance in the UK and USA since the 1970s which has made renewal and critique of CCCS work both unavoidable and urgent.

7 As Antony Easthope has quite rightly claimed; see 1991, especially chapter 5.

8 On the definition and value of this term see introduction to Lash and Friedman 1992.

9 See Toop 1984, and Steward and Garratt 1984.

10 See Redhead and McLaughlin 1985.

11 See Whittle (1994) for a description and analysis of this and other similar parts of many world cities such as Manchester, Toronto and San Francisco.

Glossary

This glossary is intended to be a brief guide to some of the concepts already in use in this field and also some of the terms specifically conceived for the current project in this book of re-presenting and analysing historical and contemporary narratives about law and popular culture.

Aesthetics of law is the art of law. An 'aesthetics of' just about every other cultural form (from music to everday life) has been developed; law, as usual, is the last bastion to fall.

Body-in-law, as opposed to subject-in-law, is a term which refers to the legal conception, and construction, of the body.

Cyber law is the (imaginary) regulatory regime for cyberspace.

Dead law is the realm of the ends of law.

Deviance means the breach of social and moral as well as legal rules.

Erotics of law is the sexualisation of law, the way in which law itself becomes desired, seduced and consumed.

Hyperlegality is the legal condition of hypermodernity, itself a term used by those who wish to stress the excesses and extremes of modernity without using the 'post' word. Like both Umberto Eco and Jean Baudrillard's notions of hyperreality, this term is meant to imply a condition more 'legal than legal', the excess of law.

Jurisprudence is conventionally the theory part of the curriculum in law schools. It is the name of courses in the theory, or philosophy, of law. As we approach the end of the millennium it is said by some commentators that we are witnessing the disappearance, or end, of law. Even those who disagree with this viewpoint recognise an increasing anxiety about the answers to traditional jurisprudential questions – what is law? what does it do? why does it do it? who does it do it to? Moreover, the search for a purely legal theory about law has attracted many critics. Theories from disciplines other than law such as literary and linguistic theory, psychoanalysis, social science, cultural studies and gender studies, as well as, more conventionally, philosophy, now figure strongly in the study of jurisprudence. Journalism, films and contemporary novels are as essential 'texts' for analysis in jurisprudence today as statutes, judgments in court, official reports and academic legal commentaries.

The last law? is always already a founding question of modernity, like panic encyclopaedists Arthur and Marilouise Kroker's idea of 'the last sex'. It is the state of no return for law.

The love of law is a term adapted from the work of French legal theorist Pierre Legendre, who refers in his work to the love of politics.

Law and order campaign is a concept used by the Centre for Contemporary Cultural Studies in the 1970s to denote the 'looping' together, by the mass media and others, of a whole series of discrete moral panics.

Moral panic was described in the 1960s as staffing the moral barricades with 'right' thinking members of society.

Panic law is meant to denote the frenzied-but-simulated state of law and justice at the end of the century, as in Jean Baudrillard's use of 'panic crash' to describe global economic stock exchange breakdowns.

Popular culture is the 'other' of high culture, but it also refers to the terrain of the deconstruction of high/low culture.

Populism is the 'popular' brought down to its lowest, market-oriented, denominator.

Post-law denotes the field created by the death of the discipline of jurisprudence.

Postmodernism is a concept which has been, misleadingly, associated with the decade of the 1980s. It has come to designate almost anything, so as to register, unfortunately, on many occasions, a sense of complete meaninglessness; in some ways this is appropriate since one of of the connotations of its use is the 'end of meaning', or the 'end of the referent'. First used in the 1950s, it became common currency in academic circles during the 1980s and 1990s across virtually every discipline from fine art to law although it has been used differently in almost every context. Importantly, it is a term that has gained prominence in cultural circles outside the academy so that, for instance, the USA has a 'postmodern pop' music chart; indeed one of the most relevant uses of postmodernism has been to designate a breaking down of the historical boundary between the academic and non-academic. The most significant barrier which postmodernism has referred to is perhaps that of high and low culture. Breakdown of this 'great divide' has been one feature pointed to by commentators on the so-called 'postmodern condition' as Jean-François Lyotard christened it, but the constant reconstruction as well as the de(con)struction of such a binary line has also been noteworthy. There is considerable disagreement, too, about whether postmodernism is a feature of an historical period after modernity, or else the ever-present founding condition (the other) of modernism itself. The idea of postmodernist style – of writing, architectural design, and so on – is that of playful, jokey, parodic self-reflexivity. If there is consensus at all about postmodernism as a concept utilisable for analysis it is in the idea of a process of postmodernisation where everything is tending to become cultural rather than economic, political or social.

Queer law denotes both the use of the legal system, legal practice and legal theory by gay and lesbian culture and the homosexualisation of legal desire.

Safe law is the legal equivalent of safe sex; it comes after the death of law, after the corpse of law has been fought over.

Transgression refers to the breaches of rules.

Translaw denotes the breaking down of the boundary between law and non-law. Inspired by an extension of the use of concepts like 'trans-political' and 'transaesthetics' by Jean Baudrillard who also saw 'transsexual' in the same light. This is the law cross-dressing, the law 'in drag'.

Bibliography

Acker, Kathy (1984) *Blood and Guts in High School, plus Two* (Pan, London)

Acker, Kathy (1986) *Don Quixote* (Paladin, London)

Acker, Kathy (1989) *Young Lust* (Pandora, London)

Allen, Richard (1992) *The Complete Richard Allen, volume one* (ST Publishing, Troon)

Allen, Richard (1993) *The Complete Richard Allen, volume two* (ST Publishing, Troon)

Arnold, Gina (1993) *Route 666: On the Road to Nirvana* (Picador, London)

Avgikos, Jon (1993) 'Cindy Sherman: Burning down the house', *Artforum*, January

Bailey, Peter (ed.) (1986) *Music Hall: The Business of Pleasure* (Open University Press, Milton Keynes)

Baker, Houston A. Jr (1993) *Black Studies, Rap and the Academy* (University of Chicago Press, London)

Baker, Nicholson (1992) *Vox* (Granta, London)

Baker, William (1979) 'The making of a working class football culture in Victorian England', *Journal of Social History*, vol. 13, no. 2

Barker, Martin (1983) 'How nasty are the nasties?', *New Society*, 10 November

Barker, Martin (1984a) *A Haunt of Fears: The Strange History of the British Horror Comics Campaign* (Pluto Press, London)

Barker, Martin (ed.) (1984b) *The Video Nasties: Freedom and Censorship in the Media* (Pluto Press, London)

Baudrillard, Jean (1988) 'Hunting Nazis and losing reality', *New Statesman*, 19 February

Baudrillard, Jean (1990) *Seduction* (Macmillan, London)

Baudrillard, Jean (1991) 'The reality gulf', *The Guardian*, 11 January

Baudrillard, Jean (1993) *Symbolic Exchange And Death* (Sage, London)

Beaumont, Peter (1993) 'Sweaty palms and safe sex', *The Observer*, 21 November

Beirne, Piers and Hunt, Alan (1988) 'Law and the constitution of Soviet society', *Law and Society Review*, vol. 22, no. 3

Beirne, Piers and Quinney, Richard (eds.) (1982) *Marxism and Law* (John Wiley, New York)

Bennett, Tony, Frith, Simon and Grossberg, Larry (eds) (1993) *Rock and Popular Music: Politics, Policies, Institutions* (Routledge, London)

Bennett, Tony, Waites, Bernard and Martin, Graham (eds) (1982) *Popular Culture: Past and Present* (Croom Helm, London)

Berubé, Michael (1992) 'Pop goes the Academy', *Village Voice Literary Supplement*, April

Bibbings, Lois and Alldridge, Peter (1993) 'Sexual expression, body alteration and the defence of consent', *Journal of Law and Society*, vol. 20, no. 3

Bottomley, Ann and Conaghan, Joanne (eds) (1993) *Feminist Theory and Legal Strategy*, special issue of *Journal of Law and Society* (Blackwell, Oxford)

Bottomley, Ann and Gibson, Susie (1987) 'Dworkin, which Dworkin?', *Journal of Law and Society*, vol. 14, no. 2

Bourdieu, Pierre (1987) 'The force of law: Towards a sociology of the juridical field', *Hastings Law Journal*, vol. 38

Bowcott, Owen and Hamilton, Sally (1993) *Beating the System: Hackers, Phreakers and Electronic Spies* (Bloomsbury, London)

Bracewell, Michael (1988) *The Crypto-Amnesia Club* (Serpent's Tail, London)

Bracewell, Michael (1991) *The Divine Concepts of Physical Beauty* (Minerva, London)

Bracewell, Michael (1992) *The Conclave* (Secker & Warburg, London)

Bracewell, Michael, Watson, Don and Edwards, Mark (1988) *The Quick End* (Fourth Estate, London)

Bresler, Fenton (1989) *Who Killed John Lennon?* (St Martin's Press, New York)

Bronfen, Elisabeth (1992) *Over Her Dead Body: Death, the Feminine Body and the Aesthetic* (Manchester University Press, Manchester)

Brown, Adam (1993) 'Football fans and civil liberties', School of Law/ Manchester Institute for Popular Culture Working Paper, the Manchester Metropolitan University

Brown, Beverley (1980) 'Private faces in public places', *I and C*, no. 7

Brown, Beverley (1981) 'A feminist interest in pornography – some modest proposals', *m/f*, nos 5 and 6

Brown, Beverley (1982) 'A curious arrangement: interview with James Ferman', *Screen*, vol. 23, no. 5

Brown, Beverley (1993) 'Troubled vision', *New Formations*, no. 19

Brown, Richard (ed.) (1984) *UK Society: Work, Urbanism and Inequality*, 2nd edition (Weidenfeld & Nicolson, London)

Buford, Bill (1991) *Among the Thugs* (Secker & Warburg, London)

Burgess, Anthony (1972) *A Clockwork Orange* (Penguin, Harmondsworth)

Cain, Maureen and Hunt, Alan (1979) *Marx and Engels on Law* (Academic Press, London)

Carey, John (1992) *The Intellectuals and the Masses: Pride and Prejudice among the Literary Intelligentsia 1880–1939* (Faber & Faber, London)

Carrington, Paul (1984) 'Of law and the river', *Journal of Legal Education*, vol. 34

Carty, Anthony (ed.) (1990) *Post-Modern Law* (Edinburgh University Press, Edinburgh)

Carty, Anthony and Mair, Jane (1990) 'Some post-modern perspectives on law and society, *Journal of Law and Society*, vol. 17, no. 4

Chapman, Robert (1992) *Selling the Sixties* (Routledge, London)

Chapman, Robert (1994) *Undercurrents* (Manchester University Press, Manchester)

Chase, Anthony (1986a) 'Toward a legal theory of popular culture', *Wisconsin Law Review*

Chase, Anthony (1986b) 'Lawyers and popular culture: A review of mass media portrayals of American attorneys', *American Bar Foundation Research Journal*, no. 2

Cheney, Deborah (1993) 'Visual rape', *Law and Critique*, vol. 4, no. 2

Chesterman, John and Lipman, Andy (1988) *The Electronic Pirates: DIY Crime of the Century* (Routledge/Comedia, London)

Chevigny, Paul (1991) *Gigs: Jazz and the Cabaret Laws in New York City* (Routledge, New York and London)

Collins, Hugh (1987) 'Roberto Unger and the critical legal studies movement', *Journal of Law and Society*, vol. 14, no. 4

Cixous, Hélène and Clément, Catherine (1987) *The Newly Born Woman* (Manchester University Press, Manchester)

Clarke, John (1991) *New Times and Old Enemies: Essays on Cultural Studies and America* (Harper Collins, London)

Cloonan, Martin (1991) 'Censorship and popular music', Institute of Popular Music Working Papers no. 1, University of Liverpool

Cloonan, Martin (1993) 'Not taking the rap – NWA get stranded on an island of realism', paper to *International Association for the Study of Popular Music* (IASPM) Conference, University of Stockton, California, 12 July

Cohen, Jeremy and Gleason, Timothy (1990) *Social Research in Communication and Law* (Sage, London)

Cohen, Stan (ed.) (1971) *Images of Deviance* (Penguin, Harmondsworth)

Cohen, Stan (1974) 'Breaking out, smashing up and the social context of aspiration', *Working Papers in Cultural Studies*, no. 5, Centre For Contemporary Cultural Studies, University of Birmingham

Cohn, Nik (1989) *Ball The Wall* (Picador, London)

Collier, Richard (1991) 'Masculinism, law and law teaching', *International Journal of Sociology of Law*, vol. 19, no. 4

Cope, Nigel (1993) 'Sony faces a test of faith', *The Independent on Sunday*, 24 October

Coombe, Rosemary (1992) 'The properties of culture and the possession of identity: Preoccupations of colonial and postcolonial societies', unpublished manuscript, University of Toronto

Cooper, Dennis (1992) *Frisk* (Serpent's Tail, London)

Cooper, Dennis (1994a) *Wrong* (Serpent's Tail, London)

Cooper, Dennis (1994b) *Closer* (Serpent's Tail, London)

Cornell, Drucilla (1992) *The Philosophy of the Limit* (Routledge, London)

Cornell, Drucilla, Rosenfeld, Michael and Carlson, David Gray (eds) (1992) *Deconstruction and the Possibility of Justice* (Routledge, London)

Corrigan, Paul (1979) *Schooling the Smash Street Kids* (Macmillan, London)

Cotterrell, Roger (1981) 'Conceptualising law: Problems and prospects of contemporary legal theory', *Economy and Society*, vol. 10, no. 3

Coupland, Douglas (1992) *Generation X* (Abacus, London)

Coupland, Douglas (1993) *Shampoo Planet* (Simon & Schuster, London)

Cousins, Mark (1987) 'Socratease', *New Formations*, no. 1

Crook, Stephen, Pakulski, Jan and Waters, Malcolm (1992) *Post-modernization: Changes in Advanced Society* (Sage, London)

Cross, Brian (1993) *It's Not About a Salary: Rap, Race and Resistance in Los Angeles* (Verso, London)

Critcher, Chas 'Football since the war' (1979), in Chas Critcher, John Clarke and Richard Johnson (eds) *Working Class Culture: Studies in History and Theory* (Hutchinson, London)

Critcher, Chas, Clarke, John and Johnson, Richard (eds) (1979) *Working Class Culture: Studies in History and Theory* (Hutchinson, London)

Cunningham, Hugh (1980) 'Review of Tony Mason, *Association Football and English Society*', *Bulletin of Society for Study of Labour History*, no. 41

Cunningham, Hugh (1982) 'Class and leisure in mid-Victorian England', in Tony Bennett, Bernard Waites and Graham Martin (eds) *Popular Culture: Past and Present* (Croom Helm, London)

Davies, Gillian and Hung, Michele (1993) *Music and Video Private Copying: An International Survey of the Problem and the Law* (Sweet & Maxwell, London)

Davies, John (1990) *Youth and the Condition of Britain* (Athlone Press, London)

Davis, Mike (1990) *City of Quartz: Excavating the Future in Los Angeles* (Verso, London)

Deleuze, Gilles and Guattari, Félix (1983) *On the Line* (Semiotext(e), New

York)

Deleuze, Gilles and Guattari, Félix (1986) *Nomadology* (Semiotext(e), New York)

Denzin, Norman (1991) *Images of Postmodern Society* (Sage, London)

Derrida, Jacques (1992) 'Force of law: "The mystical foundation of authority" ', in Drucilla Cornell, Michael Rosenfeld and David Gray Carlson (eds) *Deconstruction and the Possibility of Justice* (Routledge, London)

Dick, Leslie (1989) 'Minitel 3615', in Marsha Rowe (ed.) *Sex in the City* (Serpent's Tail, London)

Dollimore, Jonathan (1991) *Sexual Dissidence* (Oxford University Press, Oxford)

Douzinas, Costas and Warrington, Ronnie (1987) 'On the deconstruction of jurisprudence: Fin(n)is philosophiae', *Journal of Law and Society*, vol. 14, no. 1

Douzinas, Costas and Warrington, Ronnie (1991) ' "A well-founded fear of justice": Law and ethics in postmodernity', *Law and Critique*, vol. 2, no. 2

Douzinas, Costas, Warrington, Ronnie and McVeigh, Shaun (1989) 'Thrashing in the dwelling house', *Modern Law Review*, vol. 52

Douzinas, Costas, Warrington, Ronnie and McVeigh, Shaun (1991) *Postmodern Jurisprudence* (Routledge, London)

Duxbury, Neil (1989) 'Exploring legal tradition: Pyschoanalytical theory and Roman law in modern continental jurisprudence', *Legal Studies*, vol. 9

Duxbury, Neil (1990) 'Back to the Middle Ages', *International Journal for the Semiotics of Law*, vol. 3, no. 7

Duxbury, Neil (1991) 'Post-modern jurisprudence and its discontents', *Oxford Journal of Legal Studies*, vol. 11, no. 4

Duxbury, Neil (1994) *Patterns of American Jurisprudence* (Oxford University Press, Oxford)

Easthope, Antony (1988) *British Post-Structuralism* (Routledge, London)

Easthope, Antony (1991) *Literary into Cultural Studies* (Routledge, London)

Easthope, Antony (1993a) 'Why academics hate the modern review', unpublished paper, Faculty of Humanities and Social Science, the Manchester Metropolitan University

Easthope, Antony (1993b) 'Dogma bites man', *The Modern Review*, April–May

Easton, Susan, Howkins, Brian *et al.* (1988), *Disorder and Discipline: Popular Culture From 1550 to the Present* (Temple Smith, Aldershot)

Eckersley, Robyn (1987) 'Whither the feminist campaign? An evaluation of feminist critiques of pornography', *International Journal of Sociology of Law* vol. 15, no. 2

Edelman, Bernard (1979) *Ownership of the Image: Elements for a Marxist Theory of Law* (Routledge & Kegan Paul, London)

Edwards, Gavin (1992) 'Cyber lit', *Details*, June

Ellis, Bret Easton (1986) *Less Than Zero* (Picador, London)

Ellis, Bret Easton (1987) *The Rules of Attraction* (Picador, London)

Ellis, Bret Easton (1991) *American Psycho* (Picador, London)

Ellis, John (1980) 'Photography/pornography/art/pornography', *Screen*, vol. 21, no. 1

Elms, Robert (1989) *In Search of the Crack* (Penguin, Harmondsworth)

Fine, Bob, Young, Jock, Matthews, Roger and Lea, John (eds) (1979) *Capitalism and the Rule of Law: From Deviancy Theory to Marxism* (Hutchinson, London)

Fish, Stanley (1987) 'Critical legal studies', *Raritan*, vol. 7, no. 2

Fisher, Mark (1988) 'Hurd instinct', *New Statesman*, 5 February

Fisher, Paul (1992) 'Away with the words', *The Guardian*, 27 November

Fiske, John (1989) *Understanding Popular Culture* (Unwin Hyman, London)

Fiss, Owen (1986) 'The death of the law?', *Cornell Law Review*, vol. 72

Fitzgerald, Mike (ed.) (1981) *Crime and Society* (Routledge & Kegan Paul, London)

Fitzpatrick, Peter (1983) 'Marxism and legal pluralism', *Australian Journal of Law and Society*, vol. 1, no. 2

Fitzpatrick, Peter (ed.) (1991), *Dangerous Supplements: Renewal and Resistance in Jurisprudence* (Pluto Press, London)

Fitzpatrick, Peter (1992) *The Mythology of Modern Law* (Routledge, London)

Fitzpatrick, Peter and Hunt, Alan (eds) (1988) *Critical Legal Studies*, special issue of *Journal of Law and Society* (Blackwell, Oxford)

Foucault, Michel (1975) *The Birth of the Clinic: An Archaeology of Medical Perception* (Vintage, New York)

Foucault, Michel (1977) *Discipline and Punish: The Birth of the Prison* (Allen Lane, London)

Frank, Lisa and Smith, Paul (eds) (1993) *Madonnarama: Essays in Sex and Popular Culture* (Cleis Press, Pittsburgh)

Freeman, Alan (1981) 'Truth and mystification in legal scholarship', *Yale Law Journal*, vol. 90

Frith, Simon (1987) 'Copyright and the music business', *Popular Music*, vol. 7, no. 1

Frith, Simon (1990) 'No biz like the old biz', *The Observer*, 30 December

Frith, Simon (1992) 'End of the wedge', *New Statesman and Society*, 6 March

Frith, Simon (ed.) (1993) *Music and Copyright* (Edinburgh University Press, Edinburgh)

Frith, Simon and Savage, Jon (1993) 'Pearls and swine: The intellectuals and the mass media', *New Left Review*, no. 198

Frug, Jerry (1986) 'Henry James, Lee Marvin and the law', *New York Times Book Review*, 16 February

Frug, Mary Joe (1992) *Postmodern Legal Feminism* (Routledge, New York)

Gaines, Jane (1992) *Contested Culture: The Image, the Voice and the Law* (British Film Institute, London)

Gane, Mike (1991a) *Baudrillard – Critical and Fatal Theory* (Routledge, London)

Gane, Mike (1991b) *Baudrillard's Bestiary* (Routledge, London)

Gane, Mike (1993) *Baudrillard Live: Selected Interviews* (Routledge, London)

Gane, Mike and Johnson, Terry (1993) *Foucault's New Domains* (Routledge, London)

Gibson, William (1986) *Neuromancer* (Grafton, London)

Gibson, William (1987) *Count Zero* (Grafton, London)

Gibson, William (1988) *Burning Chrome* (Grafton, London)

Gibson, William (1989) *Mona Lisa Overdrive* (Grafton, London)

Gibson, William (1993) *Virtual Light* (Viking, London)

Gibson, William and Sterling, Bruce *The Difference Engine* (Gollancz, London)

Gilroy, Paul (1993a) *The Black Atlantic* (Verso ,London)

Gilroy, Paul (1993b) *Small Acts* (Serpent's Tail, London)

Goldberg, Adrian (1991) 'Radio pirates living in fear of raid-io', *The Observer*, 14 April

Goodrich, Peter (1986) *Reading the Law* (Blackwell, Oxford)

Goodrich, Peter (1988a) *Legal Discourse* (Macmillan, Basingstoke)

Goodrich, Peter (1988b) 'Simulation and the semiotics of law', *Textual Practice*, vol. 2, no. 2

Goodrich, Peter (1990) *Languages of Law: From Logics of Memory to Nomadic Masks* (Weidenfeld & Nicolson, London)

Goodrich, Peter and Warrington, Ronnie (1990) 'The lost temporality of law: An interview with Pierre Legendre', *Law and Critique*, vol. 1, no. 1

Hafner, Katie and Markoff, John (1993) *Cyberpunk: Outlaws and Hackers on the Computer Frontier* (Corgi, London)

Hall, Stuart (1981) 'Notes on deconstructing "the popular" ', in Raphael Samuel (ed.) *People's History and Socialist Theory* (Routledge & Kegan Paul, London)

Hall, Stuart, Jefferson, Tony, Clarke, John and Roberts, Brian (1978) *Policing The Crisis: Mugging, the State and Law and Order* (Macmillan, Basingstoke)

Hart, Herbert (1963) *Law, Liberty and Morality* (Oxford University Press, London)

Hay, Douglas, Linebaugh, Peter, Thompson, Edward (1975) *Albion's*

Fatal Tree (Allen Lane, London)

Headley, Victor (1992) *Yardie* (X Press, London)

Headley, Victor (1993) *Excess* (X Press, London)

Hebdige, Dick (1986/7) 'A report on the western front: Postmodernism and the 'politics' of style', *Block*, no. 12

Heller, Thomas (1980) 'A brief rejoinder to the discussion of CCLS', *Zeitschrift für Rechtssoziologie*, no. 1

Hirst, Paul (1988) 'Associational socialism in a pluralist state', *Journal of Law and Society*, vol. 15, no. 1

Hirst, Paul (1994) *Associative Democracy* (Polity, Cambridge)

Hirst, Paul and Jones, Phil (1987) 'The critical resources of established jurisprudence', *Journal of Law and Society*, vol. 14, no. 1

Hirst, Paul and Kingdom, Elizabeth (1979/80) 'On Edelman's *Ownership of the Image*', *Screen*, vol. 20, nos 3/4

Horwitz, Morton J. (1977) *The Transformation of American Law 1780–1860* (Clarendon Press, Oxford)

Hunt, Alan (1986) 'The new legal history', *Contemporary Crises*, vol. 10, no. 2

Hunt, Alan (1987) 'The critique of law: What is critical about critical legal theory?', *Journal of Law and Society*, vol. 14, no. 1

Hunt, Alan (1988) 'Living dangerously on the deconstructive edge', *Osgoode Hall Law Journal*, vol. 26

Hunt, Alan (1990) 'The big fear: Law confronts postmodernism', *McGill Law Journal*, vol. 35, no. 3

Hunt, Alan (1993) *Explorations in Law and Society* (Routledge, London)

Hunt, Alan (1994) *The Governance of the Consuming Passions: A History of Sumptuary Law*, forthcoming

Hunt, Alan and Wickham, Garry (1995) *On Foucault and Law* (Pluto Press, London)

Hutchinson, Allan (1988) *Dwelling on the Threshold* (Sweet & Maxwell, London)

Indiana, Gary (1994) *Rent Boy* (Serpent's Tail, London)

Jaff, Jennifer (1986) 'Law and lawyers in pop music: A reason for self-reflection, *University of Miami Law Review*, vol. 40

Jessop, Bob (1980) 'On recent marxist theories of law, the state and juridico-political ideology', *International Journal of Sociology of Law*, vol. 8, no. 4

Jones, Jack (1992) *Let Me Take You Down* (Virgin Books, London)

Journes, Claude (1982) 'The crisis of marxism and critical legal studies: A view from France', *International Journal of Sociology of Law*, vol. 10, no. 1

Kairys, David (ed.) (1982) *The Politics of Law: A Progressive Critique* (Pantheon, New York)

Katsch, Ethan (1989) *The Electronic Media and the Transformation of*

Law (Oxford University Press, Oxford)

Kettle, Martin and Hodges, Lucy (1981) *Uprising!* (Pan, London)

Kennedy, Duncan and Gabel, Peter (1984) 'Roll over Beethoven' *Stanford Law Review*, vol. 36

Kennedy, Roseanne (1992) 'Spectacular evidence: Discourses of subjectivity in the trial of John Hinckley', *Law and Critique*, vol. 3, no. 1

Korr, Charles (1978) 'West Ham United Football Club and the beginning of professional football in east London, 1895–1914', *Journal of Contemporary History*, vol. 13

Kuhn, Annette (1984) 'Public versus private: The case of indecency and obscenity', *Leisure Studies*, vol. 3

Lacan, Jacques (19787), *Ecrits* (Tavistock, London)

Lange, David (1992) 'At play in the fields of the word: Copyright and the construction of authorship in the post-literate millenium', *Law and Contemporary Problems*, vol. 55, no. 2

Lash, Scott and Friedman, Jonathan (eds) (1992) *Modernity and Identity* (Blackwell, Oxford)

Lash, Scott and Urry, John (1993) *Economies of Signs and Spaces* (Sage, London)

Lechte, John (1993) *Julia Kristeva* (Routledge, London)

Legge, Gordon (1989) *The Shoe* (Polygon, Edinburgh)

Legge, Gordon (1991) *In Between Talking about the Football* (Polygon, Edinburgh)

Leigh, David (1993) *Betrayed* (Bloomsbury, London)

Linder (with an introduction by Michael Bracewell) (1992) *Morrissey Shot* (Secker & Warburg, London)

Livingston, Duncan (1982) 'Round and round the bramble bush: From legal realism to critical legal scholarship', *Harvard Law Review*, vol. 95

Lury, Celia (1993) *Cultural Rights: Technology, Legality and Personality* (Routledge, London)

Lyotard, Jean-François (1983) 'Presenting the unpresentable', *Artforum*, October

Lyotard, Jean-François (1989) *The Differend: Phrases in Dispute* (Manchester University Press, Manchester)

McCaffery, Larry (ed.) (1991) *Storming the Reality Studio: A Casebook of Cyberpunk and Postmodern Fiction* (Duke University Press, London)

McGregor, Craig (1984) *Pop Goes The Culture* (Pluto Press, London)

McGuigan, Jim (1992) *Cultural Populism* (Routledge, London)

McHale, Brian (1993) *Constructing Postmodernism* (Routledge, London)

McInerney, Jay (1985) *Bright Lights, Big City* (Jonathan Cape, London)

McInerney, Jay (1989) *Story of My Life* (Penguin, Harmondsworth)

McInerney, Jay (1992) *Brightness Falls* (Penguin, Harmondsworth)

MacInnes, Colin (1980) *Absolute Beginners* (Allison & Busby, London)

McRobbie, Angela (1991) 'The worthy pursuit of pop', *The Guardian*, 5

December

McRobbie, Angela (1994) 'Folk devils fight back', *New Left Review*, no. 203

Malcolmson, R.W. (1982) 'Popular recreations under attack', in Tony Bennett, Brian Waites and Graham Martin (eds) *Popular Culture: Past and Present* (Croom Helm, London)

Martinez, Ruben (1992) *The Other Side* (Verso, London)

Mason, Tony (1980) *Association Football and English Society 1863–1915* (Harvester, Brighton)

Matza, David (1969) *Becoming Deviant* (Prentice-Hall, New York)

Medved, Michael (1993) *Hollywood Versus America* (Harper Collins, London)

Millar, Martin (1987) *Milk, Sulphate and Alby Starvation* (Fourth Estate, London)

Millar, Martin (1988) *Lux the Poet* (Fourth Estate, London)

Millar, Martin (1989) *Ruby and the Stone Age Diet* (Fourth Estate, London)

Millar, Martin (1991) *The Good Fairies of New York* (Fourth Estate, London)

Miller, James (1993) *The Passion of Michel Foucault* (Harper Collins, London)

Milne, Kirsty (1987) 'Porn: What do women want?,' *New Society*, 23 October

Milovanovic, Dragan (1992) *Postmodern Law and Disorder* (Deborah Charles, Liverpool)

Minda, Gary (1986) 'Phenomenology, Tina Turner and the law', *New Mexico Law Review*, vol. 16

Moi, Toril (1985) *Sexual/Textual Politics* (Methuen, London)

Moorhouse, Bert (1984) 'Professional football and working class culture', *Sociological Review*, vol. 32, no. 2

Moran, Les (1993) 'Buggery and the tradition of law', *New Formations*, no. 19

Morris, Meaghan (1988) 'Banality in cultural studies', *Block*, no. 14

Mort, Frank (1980) 'The domain of the sexual', *Screen Education*, no. 36

Mowry, Jess (1992) *Way Past Cool* (Chatto & Windus, London)

Mowry, Jess (1993) *Rats in the Tree* (Vintage, London)

Mungham, Geoff and Pearson, Geoffrey (1976) *Working Class Youth Culture* (Routledge, London)

Murakami, Ryu (1992) *Almost Transparent Blue* (Flamingo, London)

National Deviancy Conference (ed.) (1980) *Permissiveness and Control* (Macmillan, London)

New Mexico Law Review (1986), vol. 16, special issue on Teaching Law Through Rock Music

Norman, Philip (1991) *Elton John* (Hutchinson, London)

Norrie, Alan (1993) *Crime, Reason and History* (Weidenfeld & Nicolson, London)

Norris, Christopher (1988) 'Law, deconstruction and the resistance to theory', *Journal of Law and Society*, vol. 15, no. 2

Norris, Christopher (1991) *Uncritical Theory* (Lawrence & Wishart, London)

O'Hagan, Tim (1984) *The End of Law?* (Blackwell, Oxford)

Oldfield, Paul and Reynolds, Simon (1989) 'Glad to be male', *The Guardian*, 11 November

Paglia, Camille (1993) *Sex, Art and American Culture* (Penguin, Harmondsworth)

Papke, David (1991) *Narrative and the Legal Discourse: A Reader in Story Telling and the Law* (Deborah Charles, Liverpool)

Pearson, Geoffrey (1978a) 'Goths and vandals – crime in history', *Contemporary Crises*, vol. 2, no. 2

Pearson, Geoffrey (1978b) 'Leisure, popular culture and street games', *Youth in Society*, August

Pearson, Geoffrey (1983) *Hooligan: A History of Respectable Fears* (Macmillan, Basingstoke)

Petley, Julian (1984) 'A nasty tale', *Screen*, vol. 25, no. 2

Pfohl, Stephen (1992) *Death at the Parasite Cafe* (Macmillan, London)

Porter, Vincent (1979/80) 'Film copyright and Edelman's theory of law', *Screen*, vol. 20, nos 3/4

Porter, Vincent (1989) 'The Copyright, Designs and Patents Act 1988: The triumph of expediency over principle', *Journal of Law and Society*, vol. 16, no. 3

Poster, Mark (1990) *The Mode of Information* (Polity Press, Oxford)

Pratt, John and Sparks, Richard (1987) 'New voices from the ship of fools', *Contemporary Crises*, vol. 11, no. 1

Price, Richard (1992) *Clockers* (Bloomsbury, London)

Redhead, Steve (1978) 'The discrete charm of bourgeois law', *Critique: A Journal of Soviet Studies*, no. 9

Redhead, Steve (1984) 'Book review of David Kairys (ed.) *The Politics Of Law*', *International Journal of Sociology of Law*, vol. 12, no. 4

Redhead, Steve (1985) 'Book review of Bob Fine, *Democracy and The Rule of Law*', *Journal of Law and Society*, vol. 12, no, 1

Redhead, Steve (1986) 'Policing the field', *Warwick Law Working Papers*, School of Law, University of Warwick, vol. 7, no. 1

Redhead, Steve (1990) *The End-of-the-Century Party: Youth and Pop Towards 2000* (Manchester University Press, Manchester)

Redhead, Steve (1991) 'Some reflections on discourses on football hooliganism', *Sociological Review*, vol. 33, no. 3

Redhead, Steve (1993a) 'Book review of Paul Chevigny *Gigs*', *Social and Legal Studies*, vol. 2

Redhead, Steve (1993b) 'Book review of Ikuya Sato, *Kamikaze Bikers*', *Social and Legal Studies*, vol. 2

Redhead, Steve (1993c) 'The end of the end-of-the-century party', in Steve Redhead (ed.) *Rave Off: Politics and Deviance in Contemporary Youth Culture* (Avebury, Aldershot)

Redhead, Steve (ed.) (1993d) *Rave Off: Politics And Deviance In Contemporary Youth Culture* (Avebury, Aldershot)

Redhead, Steve (1994a) 'Book review of Gillian Davies and Michele Hung, *Music Video and Private Copying*', *MLR* forthcoming

Redhead, Steve (1994b) Book review of Simon Frith (ed.), *Music and Copyright*', *MLR* forthcoming

Redhead, Steve and McLaughlin, Eugene (1985) 'Soccer's style was', *New Society*, 16 August

Reed, Christopher (1993) 'Rhyme and crime and doing time', *The Observer*, 28 November

Robertson, Geoffrey (1979) *Obscenity* (Weidenfeld & Nicolson, London)

Robertson, Geoffrey (1980) 'The future of film censorship', *British Journal of Law and Society*, vol. 7, no. 1

Robertson, Geoffrey (1989) *Freedom, the Individual and the Law*, sixth edition (Penguin, Harmondsworth)

Robertson, Geoffrey and Nichol, Andrew (1992) *Media Law* (Penguin, Harmondsworth)

Robins, Dave (1984) *We Hate Humans* (Penguin, Harmondsworth)

Robins, Dave (1992) *Tarnished Vision* (Oxford University Press, Oxford)

Robins, Dave and Cohen, Phil (1978) *Knuckle Sandwich: Growing Up in the Working Class City* (Penguin, Harmondsworth)

Roiphe, Kate (1993) *The Morning After: Sex, Fear and Feminism* (Hamish Hamilton, London)

Rowe, Marsha (ed.) (1989) *Sex in the City* (Serpent's Tail, London)

Rubin, Gerry and Sugarman, David (1984) *Law, Economy and Society 1750–1914: Essays in the History of English Law* (Professional Books, Abingdon)

Samuel, Raphael (ed.) (1981) *People's History and Socialist Theory* (Routledge & Kegan Paul, London)

Santos, Boaventura de Sousa (1987) 'Law: A map of misreading. Towards a postmodern conception of law', *Journal of Law and Society*, vol. 14, no. 3

Sato, Ikuya (1991) *Kamikaze Bikers* (University of Chicago Press, London)

Saunders, David (1988) 'Copyright and the legal relations of literature', *New Formations*, no. 4

Saunders, David (1992) *Authorship and Copyright* (Routledge, London)

Savage, Jon (1993) 'March of the modes', *The Guardian*, 23 November

Sawyer, Miranda (1994) 'Johnny remember me?' *The Observer*, 13 March

Schofield, Jack (1993) 'The last word in discs', *The Guardian*, 4 March

Sharlet, Robert and Beirne, Piers (1984) 'In search of Vyshinsky: The paradox of law and terror', *International Journal of Sociology of Law*, vol. 12, no. 2

Sitbon, Guy (1988) 'Tele-orgies in the Aids era', *New Statesman*, 5 February

Smart, Barry (1993) *Postmodernity* (Routledge, London)

Smout, T. C. (ed.) (1979) *The Search for Wealth and Stability* (Macmillan, London)

Snyder, Francis and Hay, Douglas (eds) (1987) *Labour, Law and Crime: An Historical Perspective* (Tavistock, London)

Spivak, Gayatri Chakravorty (1990) (edited by Sarah Harasym) *The Post-Colonial Critic: Interviews, Strategies, Dialogues* (Routledge, London)

Stanford Law Review (1984), vol. 36, special issue on Critical Legal Studies

Stanley, Chris (1993a) 'Repression and resistance: Problems of regulation in contemporary urban culture (Part 1: Toward definition)', *International Journal of Sociology of Law*, vol. 21, no. 1

Stanley, Chris (1993b) 'The aesthetics of excess: Reading "American Psycho"', unpublished paper, University of Kent

Stanley, Chris (1993c) 'Spirits in the material world: Regulation and the symbolic laws of capital', *Working Papers in Popular Cultural Studies*, no. 3, Manchester Institute For Popular Culture, the Manchester Metropolitan University

Stanley, Chris (1993d) 'Repression and resistance: Problems of regulation in contemporary urban culture (Part II: Determining forces)', *International Journal of Sociology of Law*, vol. 21, no. 2

Stanley, Chris (1993e) 'Sins and passions', *Law and Critique*, vol. 4, no. 2

Stedman Jones, Gareth (1974) 'Working class culture and working class politics: Notes on the remaking of a working class', *Journal of Social History*, summer

Stedman Jones, Gareth (1984) *Languages of Class* (Cambridge University Press, Cambridge)

Sterling, Bruce (ed.) (1988) *Mirrorshades: The Cyberpunk Anthology* (Paladin/Grafton, London)

Sterling, Bruce (1993) *The Hacker Crackdown: Law and Disorder in the Electronic Frontier* (Viking, Harmondsworth)

Steward, Sue and Garratt, Sheryl (1984) *Signed, Sealed, Delivered: True Stories of Women in Pop* (Pluto Press, London)

Storch, Robert (1981) 'The plague of the blue locusts: Police reform and popular resistance in northern England 1840–1857', in Mike Fitzgerald (ed.) *Crime and Society* (Routledge & Kegan Paul, London)

Stuart, Alexander (1992) *Tribes* (Chatto & Windus, London)

Stuart, Alexander (1993) *The War Zone* (Vintage, London)

Sumner, Colin (1979) *Reading Ideologies: An Investigation into the Marxist Theory of Ideology and Law* (Academic Press, London)

Sumner, Colin (1990a) 'Rethinking deviance: Towards a sociology of censures', in Colin Sumner (ed.) *Censure, Politics and Criminal Justice* (Open University Press, Milton Keynes)

Sumner, Colin (ed.) (1990b) *Censure, Politics and Criminal Justice* (Open University Press, Milton Keynes)

Swedenburg, Ted (1992) 'Homies in the 'hood: rap's commodification of insubordination', *New Formations*, no. 18

Tagg, John (1980) 'Power and photography Part One: A means of surveillance: the photograph as evidence in law', *Screen Education*, no. 36

Tagg, John (1980/1) 'Power and photography Part Two: A legal reality: The photograph as property in law', *Screen Education*, no. 37

Taylor, Ian (1971) 'Soccer consciousness and soccer hooliganism', in Stan Cohen (ed.) *Images of Deviance* (Penguin, Harmondsworth)

Taylor, Ian (1981) *Law and Order: Arguments for Socialism* (Macmillan, Basingstoke)

Taylor, Ian (1987) 'Violence and video: For a social democratic perspective', *Contemporary Crises*, vol. 11, no. 2

Taylor, Ian (ed.) (1993) *Relocating Cultural Studies* (Routledge, London)

Taylor, Ian, Walton, Paul and Young, Jock (1973) *The New Criminology: For a Social Theory of Deviance* (Routledge & Kegan Paul, London)

Taylor, Ian, Walton, Paul and Young, Jock (eds) (1975) *Critical Criminology* (Routledge & Kegan Paul, London)

Taylor, Mark (1990) 'Law of desire, desire of law', *Cardozo Law Review*, vol. 11

Thompson, Edward (1975) *Whigs and Hunters* (Allen Lane, London)

Thompson, Edward (1991) *Customs in Common* (Penguin, Harmondsworth)

Toop, David (1984) *The Rap Attack* (Pluto Press, London)

Toop, David (1991) *The Rap Attack 2* (Serpent's Tail, London)

Toop, David (1992) 'Taking The Rap', *The Face*, January

Tribe, Keith (1981) 'Introduction to Neumann: Law and socialist political theory', *Economy and Society*, vol. 10, no. 3

Turner, Kay (1993) *I Dream of Madonna: Women's Dreams of the Goddess of Pop* (Thames & Hudson, London)

Tushnet, Mark (1980) 'Post-realist legal scholarship', *Journal of the Society of Public Teachers of Law*, vol. XV

Tushnet, Mark (1981) 'Legal scholarship: Its causes and cure', *Yale Law Journal*, vol. 90

Twining, William (ed.) (1989) *Learning Lawyers' Skills* (Butterworths, London)

Unger, Roberto (1983) 'The critical legal studies movement', *Harvard Law*

Review, vol. 96

Vamplew, Wray (1979) 'Ungentlemanly conduct: The control of soccer-crowd behaviour in England, 1888–1914', in T. C. Smout (ed.) *The Search for Wealth and Stability* (Macmillan, London)

Vamplew, Wray (1980) 'Sports crowd disorders in Britain 1870–1914: Causes and controls', *Journal of Sport History*, vol. 7, no. 1

Vamplew, Wray (1988) *Pay Up and Play The Game* (Cambridge University Press, Cambridge)

Vermorel, Fred and Vermorel, Judy (1985) *Starlust: The Secret Fantasies of Fans* (Comet, London)

Vermorel, Fred and Vermorel, Judy (1989) *Fandemonium: The Book of Fan Cults and Dance Crazes* (Omnibus, London)

Virilio, Paul (1991) *The Aesthetics of Disappearance* (Semiotext(e), New York)

Ward, Ian (1993) 'Law and literature', *Law and Critique*, vol. 4, no. 1

Wells, Celia, Meure, Dirk and Lacey, Nicole (1992) *Reconstructing Criminal Law* (Weidenfeld & Nicolson, London)

White, Avron Levine (1987a) 'Popular music and the law – who owns the song?', in Avron Levine White (ed.) *Lost in Music: Culture, Style and the Musical Event, Sociological Review Monograph*, no. 34

White, Avron Levine (ed.) (1987b) *Lost in Music: Culture, Style and the Musical Event, Sociological Review Monograph*, no. 34

Whittle, Stephen (ed.) (1994) *Margins of the City* (Arena, Aldershot)

Willemen, Paul (1980) 'Letter to John', *Screen*, vol. 21, no. 2

Williams, John (1991) *Into the Badlands* (Paladin, London)

Williams, John (1994) *Bloody Valentine: A Killing in Cardiff* (Harper Collins, London)

Wittstock, Melinda (1993) 'Losing faith', *The Sunday Times Magazine*, 2 October

Wright, Patrick (1992) 'Voice of the loud young turks', *The Guardian*, 9 July

Wright, Patrick (1993) 'Antique roadshow', *The Guardian*, 20 November

Yeo, Stephen and Yeo, Eileen (eds) (1981) *Popular Culture and Class Conflict 1590–1914* (Harvester, Brighton)

Young, Alison (1988) *Femininity in Dissent* (Routledge, London)

Young, Elizabeth and Caveney, Graham (1992) *Shopping in Space* (Serpent's Tail, London)

Young, Toby (1985) 'Saturday afternoon fever', *The Observer*, 9 June

Young, Toby (1993) 'Man bites dogma', *The Modern Review*, February–March

Index